This Book is Dedicated to

"The Teacher Of The Teachers," Eugene Fersen - SVETOZAR

By His Students

THE LIGHTBEARERS

INTRODUCTION

From the year 1904 until Eugene Fersen, also known as Svetozar, went into the Great Beyond on April 24, 1956, he lectured and taught the Science Of Being Teachings unceasingly throughout the United States, Canada and Hawaii. During his life, he personally taught more than 20,000 students, lectured before hundreds of thousands and reached many more who bought and studied the teaching he brought to the World. Some, after studying the lessons, went public with their own interpretations of these teachings by attempting to further advance these wisdoms. Still today, The Science Of Being Teachings are some of the most sought after books for those who are drawn to live the mastery of one's innate Divinity, as Eternal Living Beings.

This Book, "Advanced Teachings Science Of Being", contains unedited class lessons Eugene Fersen gave THE LIGHTBEARERS at the World Center from the 1940's until the Spring of 1956. Eugene's earlier lectures and teachings can be found in the already made public book "The Teacher," Vol I and the next edition Vol II will go into print as time permits.

Advanced Teachings
in
Science Of Being

By Eugene Fersen, TL, Svetozar.

Originator, Author, Teacher of

"Science Of Being"

"The Star Exercise"

"The Mental Contact"

Founder of THE LIGHTBEARERS

and known to the World as

"The Teacher Of The Teachers".

Advanced Teachings
in
Science Of Being"

Published 2011 by

THE LIGHTBEARERS PUBLISHING, Inc.

a division of

THE SCIENCE OF BEING WORLD CENTER

www.scienceofbeing.com

ISBN-13: 978-1463633042
ISBN-10: 1463633041

Advanced Teachings Of Science Of Being

Contents

Baron Eugene Fersen

Baron Eugene Fersen was the eldest son of the Grand Duchess of Russia, known as Marie Olga Alexandrovna of Russia. His mother knew before his birth that he was to be a guiding light for the people of this World; she called him "Svetozar," meaning "The Lightbearer," which Eugene began penning publicly when he volunteered in the Russian Red Cross during the war. Eugene's mother saw to it that her son had the proper teachers and education that would assist and support the Absolute Eternal Aspects of his Soul so as to fulfill his divine destiny.

We can only speak briefly about Eugene's father as those were the instructions and wishes left for us by the Elder Lightbearers. What we can share is what Eugene himself publicly spoke of when speaking of his father's side of the family, that, "he was a direct descendent of Count Axel Fersen." Eugene's grandfather, on his mother's side, was King Alexander the II of Russia. Eugene's uncle, by marriage, was Count Leo Tolstoy, the writer famously known for his renowned literary works "War and Peace," and "Anna Karenina." Tolstoy was one of Gandhi's greatest influences and friend. Eugene's half-sister was Queen Marie of Romania, the daughter-in-law of Queen Victoria of England. Eugene had a very close relationship with his half-sister Queen Marie; they spent many of their adult years together in the United States and her daughter, Princess Marie, stayed with Eugene until his death.

In 1901, Eugene came to the United States with his Mother the Grand Duchess and in 1904 began sharing his message and teachings known as "The Science Of Being." In 1906 through 1921 the Baron Eugene was investigated by the U.S. Government's Justice Department as a 'Possible Radical.' Eugene was a Russian Royal teaching what the U.S. Government termed as 'radical religious thinking;' this investigation accelerated during World War I, 1914-1918. In late September of 1921, the United States Government closed their investigation on Eugene. By early 1923, the U.S. Government allowed Eugene to become an American Citizen and granted him free reign to publish, through the American Press, his already acclaimed lessons, 'The Science Of Being'.

Before Eugene was sworn in as an American, the proceeding judge cautioned the Baron Eugene, that once he took the oath to become an American Citizen, he would no longer be able to take the Royal Throne he was born to empire. Eugene knew his purpose and mission wasn't to reign over Humankind but to assume a modest seat of service for the spiritual progress of 'All,' Humanity.

The Baron Eugene Fersen was intimately associated with the world's most eminent teachers, scientists and philosophers; some of his most profound personal teachers were from the lineage of the Great Magi. In the late 1800s through the mid 1900s Eugene taught or influenced many of the great teachers who are now responsible for the greatest-human-healing-potential movement of our time: Charles Haanel (The Master Key System,) Dr. Hotema, Elizabeth Towne (Publisher of Nautilus Magazine,) Wallace D. Wattles (The Science Of Getting Rich and The Science Of Being Well,) Edgar Cayce, Annie Besant (Translator of the Bhagavad Gita, Theosophist, and Leader of Woman's Rights,) Huna Max Freedom Long (great teacher of the Huna ways and teacher to the founders of the Course In Miracles,) Charles Fillmore (Founder of The Unity Church,) Samuel Clemens (author of Mark Twain), William Walker Atkinson (one of the three Initiates of the Kybalion,) Nikola Tesla, Manly P. Hall, Jon Peniel, and the list goes on. Rudolph Steiner himself was touched by Eugene's teachings and had him as a guest speaker/teacher in the Steiner Schools whenever Eugene could be available.

The Baron Eugene Fersen taught anyone who had a genuine interest in the 'Science Of Being,' the Truth, and the path to spiritual and human liberation. At the time of his parting into the Great Beyond in 1956, the Baron Eugene Fersen had personally instructed well over 20,000 students and at that time more than 100,000 people worldwide had read or been exposed to the teachings of the 'Science Of Being.' These teachings today are regarded as some of the most inspiring literary works in the study and education of Quantum Science, Spiritual Science, Human Enlightenment and the study of the Soul.

The Baron Eugene, came to share with us the Truth of these wisdoms with the hope that Humankind would free themselves from the myths that held them hostage and bound them to an un-liberated existence here on Earth. He shared that there are still vast amounts of profound wisdoms that remain veiled from Humankind because the un-liberated subconsciousness mind had become resistant to comprehending those truths. He knew and had faith that when humans enlightened their bodies, minds and raised their Spiritual Vibrations as they lived in the physical world, more would be revealed to them.

Baron Eugene Fersen, 'The Teacher of the Teachers,' gave to the world with an open heart, mind and spirit these profound truths, his life's purpose, and his 'seat-of-power and privilege.' He believed deeply that all of Humankind regardless of class, creed, gender or difference should have access to these great wisdoms that were once only privy to the rich and powerful. He acquired, as he lived, the manna all alchemists were truly looking for, the peak of spiritual attainment, whereas one's body, mind and spirit reaches Its highest vibrational aspects of spiritual evolution in physical form; whereas Spirit matter becomes one with Its pure Soul.

The Baron Eugene Fersen's life purpose was to bring to Humankind the lost principles of our first Primal Ancestors so as to assist Humanity to complete their tasks to awaken, to the 'All,' knowing and discovering the latent scintillating star that resides within each of their Absolute Eternal Souls.

The Relative Heavenly Eternal Realm
Know Thyself And Thou Shalt Know All

When Eternal Soul, through Its use of Spiritual Energy benevolently chose Mind-fully to separate from the Graces of the Heavenly Eternal Realm and Its Superconscious oneness with the Great Principle – The Creator, a new "Bio-Neuro-Spiritual World," and a Solar System to support Its inhabitants, were artfully born in this Universe. These new systems were solely encrypted with all the wisdoms and Life Force Energy to support and nourish Soul's sophisticated yet subastral creation, as long as the skills to decipher and access the Nature of their algorithm designs was preserved within the lineage of Soul's Spirit as it lived in this new World.

Hence, the Souls that came forth from this phenomenon emerged in elemental form- crystallized vibrations as separate aspects of the One true Nature of their prior Divinity and began creating substance to support their mission; their purpose in a now material world. These now Earth bound Souls, backed with the help of "The Great Law;" and with the hope that Supreme Harmony would reign in their new world, believed they would achieve and prevail at their Earthly purpose and task to create a Heavenly Eternal physical world.

Souls that once inhabited the Eternal Realm aspired to experience the spiritual alchemy of a "Bio-Neuro-Spiritual World." There encoded and housed in these newly formed matter-bound Heavenly bodies, was their "Eternal Supreme DNA." These Souls outfitted with the profound aspects and abilities of their Supreme Eternal Nature encoded in their human DNA; backed with their divine intent, began to re-create collaboratively the nature of their divinity in this new world. Souls in the Eternal Heavenly Realm, the Realm of Harmony, never knew fear, aging, death, hunger, poverty, physical suffering, pride, hatred or lacked the power of Love - attraction.

In the Eternal Realm, the Universal Laws operate as one law and those laws are infinitely governed by three immutable forces called "The Great Law." But, here in the Relative Realm there would be altered aspects to the Universal Laws in relation to Soul made manifest in a physical world: First - that Soul, through the Eternal Force of Its Spiritual Energy would have access to and the use of Its Supreme Aspects of Creative Force so as to create what It desired at Mind's free will. Secondly - in the Earthly Realm, the Universal Laws that were once inseparable Eternal Forces became separate Eternal Aspects of their once united Heavenly Nature. And Thirdly – If Soul's Spiritual Vibrations are not in tune with the Supreme Nature of Eternal Harmony before reentering the Eternal Realm, then by way of the governing aspects of the Law of Attraction and the supporting virtues of the Law of Evolution, Soul would be drawn again to incarnate solely with the intent to liberate Its Spiritual Vibrations from the magnetic rhythms that bind Soul's Spirit to the material plane of existence.

As these Souls took root in the Relative Realm, they began to manifest and create life without a conscious understanding about there being a physical separation from their Divine Nature, the Universal Laws, their Creator or one from another. But this would change throughout their evolution. While these Souls forged and carved out their new world they simultaneously created energy-dynamic relationship with all that they had created through the Laws of Cause and Effect, the Law of Rhythm and the non-negotiable Immutable Laws of the Universe. They began to Love and worship the perishable instead of the Eternal and created Laws from untruths instead of embracing the Laws that once brought them salvation, preservation and Liberation.

Most Scintillating as a Brilliant Star was to be the prevailing aspect of the Eternal Mind made manifest. But, throughout time on the physical plane of existence, mind was deceived by Its own power and misplaced Its benevolent Truth, Divinity and Eternal plan. Spirit reuniting with Its Eternal Twin, Soul and awakening to Its Divinity, has become Humanity's age old quest for the unknown. To "Know Thy Self," as a living manifestation of the All Powerful living force of the Eternal and to make this Wisdom Law (Truth,) once again has now become Humankind's foremost aim and their mind's greatest challenge.

The choice to transcend the Physical Realm and return to the ways of the Eternal was a path never lost; there at the bridge to the Infinite is a gate that swings freely and at will between these Worlds. Still today it seems for Humankind, that there is yet a long road to travel before Liberation is realized and materialized on the Earth Plane; but with the help of The Great Law and the merciful aspects of the Law of Evolution and its loyal ally Hope, Humanity will prevail.

As more Light is shed on the Wisdom that is before me in the rare archival documents written and taught by The Teacher of The Teachers, Eugene Fersen, I have been shown that there has always been a proper scientific and spiritual term for our Earthly Realm. It seems over time, it has escaped Humanity's grasp to embrace and be-hold. Now that we have entered the "Sixth Race of Humanity," the race of the "New Age," where much Truth has been laid before the feet of Humankind, it is now suitable and a moral obligation to reveal what has been shown to me and from this day forward address this realm by its proper and All Powerful name, "The Relative Heavenly Eternal Realm."

I have come to fully understand, as I embrace my purpose, my Destiny as Acting Head Lightbearer for the Science of Being ~ Lightbearers World Center the "Why," I was chosen to hold the space of the Seat Of Inspiration for this generation – that I would in my time, carry out the most important virtue of its duty - to put forth the unpublished Wisdom left for me in print without reservation or withholding. I have intimately come to know the Baron Eugene Fersen by reading and being with his work since I was a child. I was taught by his words and by the actions of my Lightbearer Family, the deeper meaning of Truth and shown the depths and tenacity of Love. While Eugene lived, he eradicated the forces of the rhythms that bound the human Spirit to the polarities found in this World.

As I continue to unfold and put into print the rare archival information, I know without a shadow of a doubt that these teachings and other wisdoms still not yet revealed were never to be kept secret from Humanity. These unprincipled actions have created and supported a consumptive deterioration in the Evolution of Humankind. Eugene never referred to the Science Of Being Teachings as Esoteric or attributed their wisdoms to any one philosopher or teacher except for the Great Principle and for very good reason, as those statements worked against the Law – Truth. He taught The Science Of Being teachings freely, whenever one was ready to learn and wherever he was given the opportunity to speak about them; he did so with a devotional determination that was unwavering, until he went into the Great Beyond.

The Seat of Inspiration, the seat Eugene Fersen sat in, has been an immeasurable seat for me to behold; I do my best to honor my Destiny and purpose in a good way; with a great respect for All Life. For as many thumbprints that exist in human form and blueprints found in Nature, that's how many religions reside on our planet. The path that each Spirit must live to become One again with Its Eternal Soul and Its Creator is as unique as Its thumbprint. As I honor the incomparable you, I also honor your personal relationship with the Eternal and the path – the "Why," you must live so as to unearth your Divine Destiny and return to the gate that swings freely and at will between the worlds.

Sherrie ~ Love, blessed Soul
May the flame of Love always
light your way!
Laura Taylor-Jensen AHL

A Note For The Reader

"The Advanced Teaching in Science Of Being Vol. I," is comprised of unedited advanced teachings and class lectures given by the Baron Eugene Fersen in the early 1900's. Grammar, word usage, some scientific topics and the use of generalized gender pronouns found in these writings, are befitting and reflect the period before, during and post WWII academic style of that era.

The Lightbearers ~ Science Of Being ~ World Center's, Acting Head Lightbearers, would like to offer some insight to what Eugene meant when he used the following terms. We feel confident that these are accurate and proper interpretations. When Eugene Fersen used the term man or men in general it denotes Humanity or Humankind unless he was speaking about men as a gender type. The term problem connotes an opportunity or a task. When Eugene referred to the Father as the Great Architect or Great Principle in that statement, was included the sum of Its whole parts, meaning the Eternal Mother-Architect-Principle aspect was included and present. And Truth meant Law.

Lesson One

THERE IS A GOD

Svetozar read from his book, "Is There a God?" paraphrasing the first paragraph of page 215. *Following is the paragraph:*

"Self-existent, without Beginning and without End, including in Itself all Infinity of Space and Eternity of Time, creating everything, constituting everything, governing, sustaining, pervading and containing everything, that from which all things come into existence and to which all things return, the Source of all Powers, all Forces, all Laws, yet at the same time the fulfillment of them also, their expression in accomplished results, Immovable in the midst of Eternal Motion, Immutable in the midst of Unceasing Change, the Foundation of Matter, the Substance of Mind, the Essence of Spirit, the Life of the Universe, Its Creating Intelligence, Its Great Law from which all the specific Laws emerge, Its Universal Power of Attraction, binding all into a coherent whole – this is Universal Life Energy, the Supreme Ruling Power of the Universe."

If we understood this paragraph, we would solve the problem of our life. The words have an effect on us, only as we realize they are true. We should have an inner feeling of the Truth.

Self-existent, without Beginning and without End. This is the most difficult proposition ever presented to understand, and will remain so, because the human mind cannot grasp the Limitless. Because we cannot grasp It physically, mentally or emotionally, it is self-evident that It must grasp us. In the words Beginning and End, is all we want – security, protection, etc.

Infinity of Space and Eternity of Time. The reason we cannot grasp the Fourth Dimension is because it grasps us. In it, we find an explanation of Infinity and Eternity. As the Fourth Dimension surrounds us, so also the Eternal surrounds us. Every part of our body, not only the physical, is surrounded by the Fourth Dimension. Even atoms and electron are surrounded by it. In the mathematical number 4, we have included the numbers 3 and 2 and 1. The Fourth Dimension is, for the time being, symbolical of that Greater Thing, Eternity, which we will never understand. We can understand the past, which is behind us.

Creating Everything. This Universe is in process of continual building and continually reverting back to its beginning. Everything created by Man has a beginning and an end. In the eternal scheme there is no beginning or end. Infinity and Eternity must be harmonious. In Infinity and Eternity, which is the Eternal Itself, there is a harmonious fact, not just a material fact, but mental and spiritual also. The spiritual fact precedes everything. The visible or manifest is just as much a part of the Eternal as is the Power out of which it was created. It becomes a living unit.

Constituting Everything. The next activity of the Eternal is to constitute, to put in its proper place everything – the large and the small and the intermediate. In this we see the functioning of Universal Intelligence, or Mind, doing Its work. Until now only life was at work. In this proper place, we are free to move in our own orbit. In mathematics each number must remain its own self. Each number must be loyal to its own true self. The same must be true in an expanding way with everything else, we included. There is the possibility of unfolding and bringing out details, but the fundamental activities must be the same.

Governing Everything. The Laws of Nature, or God, govern everything. If we use them we get the most and best out of life. If we think we can force Nature, the result is destructive. We will learn through hard knocks, that the Law cannot be broken. We will be impaired, but not the Law. It is like trying to break through a Spiritual Wall which cannot be broken.

Sustaining Everything. There must be continual support of what was created, constituted and governed. The Universe long ago would have collapsed without this sustaining Power. The Eternal, in sustaining Its own Creation, sustains Itself. It sustains Harmony. This Power sustains us. It is a part of our eternal character. We should cultivate It. Hope is a ray of this Power.

Pervading and Containing Everything. We are in the Eternal and It is in us. It contains our physical, mental and emotional bodies, and our activities. It is That from which all things come and to which all things return. Everything in Creation comes from the depth of the Eternal Itself. Everything emerges, fulfills its mission, and returns back to the Original Source. The Power of protection and guidance is with us from beginning to end. We are worth more, in a way, than a blade of grass. Protection increases with the increase of the value of an individual. No matter in what activity of life we have to function, the Law works. We all must fulfill the Law. We are representatives of the Law and must try to fulfill It.

Lesson Two

UNIVERSAL ENERGY

The Eternal has two aspects – Cause and Effect.

Every cause is usually unmanifest. The cause is the unknown. The effect is the eternally known, the solution of the riddle.

The Eternal is both the Law and the Fulfillment of the Law. The Law is the unmanifest. The Fulfillment is the manifest.

We only know about the Eternal when It manifests. The Eternal is Eternity and Infinity. Universal Energy is the first manifestation of the Eternal. It is like a stream flowing from the Eternal.

Characteristics of Universal Energy. It is self- existing, because it was forever in existence as a manifestation. It has its existence in the Eternal. It is like the breath of the Eternal, without beginning and without end.

It is omnipresent because it is everywhere at all times and in full power. It is the shoreless ocean of All Power. It has no limitation of shore as an ocean has. It cannot lose Itself in infinity of space, because It is space and It determines infinity of space. It also makes eternity, because, not being able to lose Itself in Itself, It must abide forever in Itself.

It is Supreme Intelligence manifest as effect, the cause being in the Eternal.

In Universal Energy there is Law.

Harmony manifests through the Law of Attraction.

Everything visible is created by that invisible Power. It has the power of creation, also the quality of constitution. The Power of Attraction preserves everything.

When Universal Energy did creative work, which never had a beginning and will never have an end, because the work has always been continuous, It did not express all of Itself in Its creation. If so, there could be no circulation. The greater volume of It still remains as a Power. It is like ice on the surface of a river. Beneath the ice, the river still flows. The ice melts and returns back to the water. It is the same with Universal Energy. Part of it is invisible and part of the visible Universe. There is a circulation between the visible Universe and its original Source.

Universal Energy pervades everything. It is outside and inside of Its own creation. Creation is like a living water condensed into a different aspect, as ice on a river.

Man is a compound manifestation into an individualized aspect of that Power. Man is only a tiny molecule in the whole block of creation. Man, in a general way, means being on all planets. We do not know in what stage of development the inhabitants of other planets are. Through reasoning, we know those beings must have the same fundamental characteristics as we have.

Lesson Three

THE FOURSQUARE PRINCIPLE

The FourSquare, or Life, Mind, Truth and Love, can never be altered. All aspects of It are in Universal Energy. At the same time, Universal Energy is emanating out of the FourSquare into the Universe.

Man, who is the individualized projection of the Great Principle into Its own Eternal Substance, has within himself the four aspects of the FourSquare, the same as in all the Universe. The four fundamentals are to some extent manifested in all humans. As a tree unfolds in its growth, so humans manifest more and more of the four aspects as they progress in their evolution.

Humans are not as full of life as are animals. We have more of it when we are born. Life ebbs as we grow older. This is a contradiction of the rest of Nature. In all of Nature, when the limit is reached, physical dissolution takes place. Science and religion teach that humans are limited to three score and ten years. There is no reason for this teaching.

Mind is not supposed to ever stop growing. It is ageless. It cannot manifest except through a body. As the body grows old we become more and more inefficient. The brain becomes an inefficient channel for mind manifestation. The mind expresses itself less and less. It is like the lens of an eye. If it is impaired, the inherent sight of the brain cannot manifest.

Our own body interferes with the manifestation of our mind. People could live thousands of years if it were not for the inadequate and improper manifestation of our minds.

Fear is the main cause of all our troubles. Wild animals will not attack persons who have no fear of them. Fear and doubt upset the whole functioning of our body.

Mind interferes with both Life and Truth. It also interferes with Love, which should coordinate everything in our body. Love is always optimistic. Optimism is one of the manifestations of Love. People would have more optimism if they had more Love.

Of all the aspects of the FourSquare Principle, Love is the leader and the most important. Human love is but a feeble reflection of that Supreme Power.

The more we reflect Love, the more we are in tune with It. If the sun shines on dirty water, the sun is not reflected by the water. Universal Energy shines on all of us, but unless we reflect It, we are not benefitted by It.

There are invisible rays beyond the rays of Love, but they are not supposed to be reflected by us. They are beyond us.

We are a mirror, created by the Eternal, to reflect all rays. The mirror is covered by dust originated from our own mind. Mind started a friction with the Great Law. The Great Law was not injured by this friction. The more friction, the more dust was laid down on the perfect mirror of our being. The mirror remained undamaged, but the dust interfered with its fulfilling its mission.

The joy which the original reflection caused is beyond human conception. It is bliss. But there is a thick, thick covering of dust which is thickening more and more for most people.

When the accumulation of dust became so thick people could no longer stand it, the people broke the crust, and it was like an earthquake on the mental plane. Humanity emerged a little better. The dust was partly destroyed and Humanity, for a while, was relieved. However, some dust closest to the mirror still remained clinging to it. The traits were lessened, but not eliminated. The mirror was still obscured.

Then again that fight continued, but on a larger scale. There was another layer of dust, another mental earthquake, another cycle ended, and another expansion.

The Sixth Cycle is now reached. The mirror of our being now is covered with a thinner, but more solidified coat of dust than ever before. It is so opaque that we can now hardly reflect at all the beauties of Love.

The mirror covered with dust is only an effect. The cause is back of it and is invisible. The effect would collapse without the cause. The cause is always invisible. The effect is always visible on the material plane. If the cause is removed, the effect disintegrates.

The invisible rays of Love meet within ourselves the same power which is buried there. These rays of Love, which contain the other three aspects of the FourSquare, penetrate through the dust and reach the Invisible within. Then the Invisible within makes an effort to arouse Itself, to come to the surface and manifest Itself. It cleans the mirror from underneath. This cleansing has been going on since the first precipitation of mental dust.

This is known to us as the great battle between Good and Evil. Mind wants to rule, rather than to turn to the Father. Some religions and philosophies think this fight will go on forever. It must come to an end some day, with victory for Good. At the end of the Sixth Cycle, Evil will explode in a most conspicuous way.

The Sixth Cycle, which we are now entering, will seem to be a Cycle of Peace. It will really be a cycle of the greatest underground work of Evil. It will be very deceptive. The last battle will be at the end of the Sixth Cycle. There will be no such battle now at the end of the Fifth Cycle, but our present condition is very important to us now. It is the battle between the good deep within us and the covering of evil.

Most of us start life in the right direction, but we are conceited and stubborn, and go farther and farther in the wrong direction. The ultimate aim is perfection, but there are many steps to it. We go in circles. We discover that we have spent the whole of our life and are back where we started. Our incarnation would produce almost no effect, if it were not for the Law of Evolution. Pride caused our downfall and stubbornness is the chief human aspect of pride. We all need to clean the mirror of our mind.

Mind has failed. The present world situation proves it. So we have to turn to Love. We cannot make Humanity as a whole see this, but we must try to realize it ourselves and face it.

We are each a mirror, created by the hands of the Eternal. In this mirror, created figuratively of diamonds, is reflected all the beauty of Creation. A diamond reflecting the sunlight is not only brilliant, but it reflects warmth (Love). The diamonds out of which we ourselves are made are the finest Mother Nature created, but we are covered with dust, because we knowingly or unknowingly are fighting the Eternal. Sometimes we see the real beauty and harmony underneath.

Cleaning this mirror is the greatest problem Humanity ever undertook. It started eons ago. The human mind is an exile from Harmony. This is symbolically described in the legend, "The Dream." A rescuer (Love) comes from the Realm of Harmony. That power alone can help us, because that Power can penetrate through every barrier.

If we wash our hands with soap, we think they are clean, but tests by scientists prove that they are not clean. The cleaning is only superficial. Usually we try to dust our mirror by ordinary means, but this is only superficial. If we use the Eternal Power of the Infinite, then from within us that Power awakens our Higher Self, which is the pure mirror of diamonds.

This work which has been carried on through countless generations is shortened by Love. If done by ourselves alone, the burden is almost unbearable at times. If we realize that the Power is working for us, then we feel relieved and the burden becomes almost a pleasure.

The greatest interferences in this work are our own shortcomings. Fear is the worst of all. Next to fear, we have to combat within us hatred, which expresses itself through countless channels. It causes humans to lose the direction they should normally take. Jealousy, doubt, suspicion are all stumbling stones in our journey. All of these shortcomings are the very handicaps which interfere with the cleaning of our mirror.

We must understand the situation, and use determination. We are facing the world under most ominous forebodings of a very dark future. We should try to harmonize ourselves and develop the love nature within us. The situation will become so serious that millions will want to die, but will not be able to do so. We should never give up when we think we are in the right direction.

Lesson Four

THE FIVE STATEMENTS OF BEING

LIFE. Human life is a part of Eternal Life. The sum of all lives in all of Nature form the One Life, which we call the Eternal. If we realize this, we feel that we are not lonely, not helpless, not hopeless.

INTELLIGENCE. The sum of all intelligence in the Universe is ruling everything for all time. Our intelligence is a ray of that Universal Mind, the Creator of everything. When we feel this, it helps us to climb the Mountain of Improvement, the steep hills ahead of us. We by and by work out our problems in a satisfactory way.

TRUTH. The sum of all truths is the Eternal Omnipotent Law. We must be sincere. Honesty is an aspect of sincerity. If we are disappointed in all human beings, there is that One Power which is the Eternal Law. This law never fails us, but we must be open to It. It is Universal Truth. We must not turn our backs to It.

We are not sincere. We are exhibitionists. Nature does not exhibit itself. It *is*. There is an inner naturalness in ourselves, but we do not express it.

LOVE. The sum of all loves in the Universe is the Eternal Love. True love is a sense of peace and harmony. When we have this feeling, we know we are contacting the Universal Law of Attraction.

BEING. The sum of all beings is our Eternal-Father-Mother. Beings include body, mind and soul, the three in one. This refers to everything in the Universe that has a body.

The Father functions through us. He sees, hears, touches, tastes, smells, all through us. Through these means the Father is communing with His own Creation. When we love in the right way, it is the Father loving through us. It is the same with telling the truth. When we are energetic, it is not our own vitality. It is the energy of the Father. He is the fulfillment of His own Law.

Why are our senses so imperfect, if the Eternal is functioning through us? Our own mind has interfered. It is so opaque that the channels of expression are obstructed. We have come to one of the darkest ages of Humanity because of our human minds. We must realize that the Powers of the Eternal are working for us.

The Five Statements of Being are more to be felt than realized mentally. The more we try to pierce the veil, the thinner the veil gets. Finally we will, in time, entirely wear out the veil. We should be in touch as much as possible with our own Higher Self and by and by we will be in touch with the Eternal.

If the human mind persists in its obstinacy, it will win its own victory, to its own defeat. Love is the warming element in a cold wintry night. It lights a candle in the window which beckons us to come in and find shelter.

Lesson Five

THE QUEST FOR THE SOUL

The body does not have a soul. The soul owns the body. The soul is an individualized, compound projection of the Infinite, containing all the qualities of the Eternal. Rays of the Universal Soul are continually proceeding and embodying themselves into a living soul, which is indissolubly connected with the Universal Soul. Our own mind obscures this fact. The problem of seeking and finding the Soul is, in a way, the most important problem in a person's life.

The Soul is the center of our existence. It is the point from which our FourSquare of Life, Mind, Truth, and Love is proceeding.

We may ask if there is a center from which these radiations come, why do we not get the radiations as we should? If we judge our human existence, it seems there is no harmonious cause back of it. But there is a screen between the cause and the effect. In most people, it is more than a screen; it is a wall.

Universal Energy passes through a wall. So it is with our own soul, no matter how heavy the wall of our human mind. If this were not so, we could not live at all. The electron could not exist if there was no center of harmonious energy. The Universal Soul is probably much more clearly manifested in the electron than it is in other beings as they rise up in the scale of Evolution. This should be the reverse. Originally it was so, and it will be so again some time. Humans have the heaviest and thickest wall of mind.

When we speak about our head, it is not the material head that is so important. The head is only the manifestation of the Soul for seeing, hearing, etc. When we think, it is our Soul which thinks. When we perceive the world through any of the other channels, it is the Soul that does it.

The Soul is at all times in communication with the outside. It is not satisfied to live the life of an introvert.

No cause, whatever it may be, can remain unmanifested. Sometimes, it takes a long time before manifestation is noticed, due to impediments of many kinds. Whenever we feel emotions or other feelings, it is only the Soul expressing as much as we can feel through the heavy wall of mind.

Not one part of our body is a part of our human body only. It is a part of the

Soul. The body is a temple in which the Soul resides. Therefore we must take good care of the body. Everything that is harmonious to us on this earth plane, such as food, air, etc., is beneficial to the Soul.

We know but little of harmony because our mind interferes with the Soul. Our mind is continually placing such difficulties in our path that we could not progress if it were not for the greater power of the Soul. If we could raise our consciousness to the mental plane, we could find the answer to the problem, "Is there a soul?"

Everyone wants to be healthy, to be happy, to be financially independent, but we do not start in the right direction. We continually start wrong causes. Like produces like. Virtue is not its own reward. Virtue means to do the right thing. We do not need the approval of other people. We do not need their advice. We can find the advice within our own selves. Something within us tells us the advice is right. We must learn to trust our own selves. Then, by and by, we will learn to trust those others who are trustworthy.

Real happiness is within us and cannot be upset by outside conditions. The human mind calls a crooked line (disharmony) straight. As long as this is so, the human mind can never see a straight line. We can never see our Higher Self at the end of a crooked line. If we do not see the ultimate harmony which our being demands, we fail to reach our goal. It really pays to be as straight as possible. It is the straight and narrow path.

Carelessness is one of the worst human shortcomings. This is true not only in what we do, but also in what we think. We should listen properly to what people say. We should see properly what is around us. Each quality is a straight line. Each shortcoming is a crooked line.

Our mind is more impressed by crookedness. An active person physically, mentally and emotionally annoys us, due to our own subconsciousness.

We are fundamentally, and always will be, beings greatly affected emotionally, because back of right emotions is Love. Mind should analyze the emotions, and, like a stream of living water, direct them in the right channel. The whole situation can be resolved and analyzed by the FourSquare Principle.

At the end of the straight line is the Soul, the lighthouse or beacon, where light is thrown upon us. The voice of Intuition is the proof that we are a Soul. We should follow intuition. Animals and plants use it. It originates from the Source where there are no mistakes. Life should be directed by the Soul, and not by the crooked, limited human mind. We will then find we are the Soul.

What is Man? Man is naturally the most important item for us to investigate because we are what are called men. Man is the body, mind and Soul. Man is an individualized projection of the Eternal into Its own Eternal Substance – a projection of Spirit into Its own Matter. Man is the Witness of the Absolute.

If there were no human beings, we would not know there is an Eternal Father, Who is the Loving Mother. This would not alter the existence of the Eternal. Our recognition of a fact is useful to us but not to the fact. It benefits only those who do the recognizing. Nothing can in any way affect or alter the fact of the Eternal.

It is important to us that we should recognize our own existence. We are the only beings on earth that can recognize their own existence. We recognize our physical body, our mind and our emotional nature, which is the best we can perceive and know of our Soul. This gives us an almost limitless aspect of everything around us. Animals do not have the power of inner analysis and perspective. They do not have the gift of analysis of things.

We have to be aware of things on the physical, mental and emotional planes. This is the only way that we can grow, that we can lift ourselves above animals. We can look back into the past, and can look forward into the future. Here again we are different from the rest of Nature.

Animals love, hate, are jealous, are loyal, but they are not aware of the deeper meaning of these emotional manifestations. We are far ahead of animals in our awareness on the emotional plane.

All of these things can be developed in humans, but not in animals. Humans can raise their rate of vibration and rise above things. Animals may perhaps do this, but they are not aware of what they are doing.

Humans usually walk through life unconscious of what Life is and how it should be approached. Yet man is an image and likeness of the Eternal. Occasionally, for a short time, the divine spark within us comes out. This gives us hope that the divine spark within us cannot be extinguished. The aspect of man which we now know is not the Real Man.

To perceive the divine spark and to know that we can fan it into a flame is very important. If there are ashes, there must have been a fire. Often among the ashes can still be found a little spark. Among mental ashes there is always the divine spark lingering. It is the flame that was within us when we were born, and is with us when we die, and then still continues. How do we know this? Because it is Life. It is indestructible, therefore eternal. That spark is, in reality, Life, the full expression of which we perceive only a glimpse in Man.

Human beings are mortal. They will perceive, by and by, underneath the mortal, the Immortal. We should fan the spark into a growing and growing flame. The flame will do its own work. It is our duty to fan it. When we fan the better qualities, those qualities will by and by automatically consume our shortcomings. No human being who is lazy will ever succeed. We have to use our whole life to fan all our fine qualities. To think of doing it is not satisfactory. We must manifest it on the physical plane. Love is always manifested in some way, such as a smile or an expression on the face.

We should try to realize within us what we really are – an individualized spark of the Eternal, indissolubly connected with the Eternal, having all the latent Powers to manifest the best that is to be manifested. Fear is our worst enemy. We are living now in most crucial times, probably the most crucial ever.

What is it to know the game of life? It is to understand life as life really is. We should not live under illusions. Wishing alone will never give us what we want. We must make a reality of our wishes. In that way we will, by and by, become a real man and not a wish man.

We must be aware of the fine spiritual qualities within us, and then bring them out. That is our duty. We should express as Mother Nature expresses. We must do it. In view of present day conditions, we must put all of our efforts in that direction. Time is short, but not too short.

Man is the Son of the Eternal. The more this is impregnated into our mind, the more it is thrown into our subconsciousness, the better we will achieve our unfoldment on this earth. Man is a compound projection of all the qualities and powers of the Eternal.

We should try to let our soul inspire us, to send into us its qualities. There is no limit to the end of these qualities. We must try to let them penetrate into our bodies. If our bodies would be well equipped in this way, no disease would affect them.

The same concerns our mind. If it is as clear, wise and loving as it should be, we would know of the wrong things of earth, but we would not be affected by them.

The same is true of the emotions. The right emotions should be the outgrowth of our Soul. Wrong emotions should be controlled. We have a mind – the power of discrimination- to do this.

As long as we are determined to fight to the end in the right direction, we are bound to succeed. When we lose faith, we lose everything.

There is now going on in our own minds a raging battle between superconsciousness and subconsciousness. Subconsciousness is going to win the battle with most people. What shall we do to win the battle? One thing which is upsetting our fight in this direction more than anything else is procrastination. If we do not do the right thing now we upset the whole chain of cause and effect. We start a chain reaction in the wrong direction. One of the secrets of harmonious functioning on earth is to do things now.

If we do the right thing we have on our side the whole of the right of the Universe. When we seem to stand alone, we are not alone. The Eternal is always around us. We are His children.

Just as our body is the product of our soul, so is also our mind. We should learn to keep it under our own government.

No one of us can take it easy now, in the present condition of the world. We have to put all the energy we can in the right direction. Then we will have more energy to use to enjoy ourselves in the right direction.

Mentally lazy people can never enjoy life. Today we are facing the situation where we must use as much judgment and as much wisdom as possible. We must master our mental armaments. We must say, "No matter what happens, there is a way out." This is wisdom.

In view of the present condition - the battle within our subconsciousness – we must learn to be optimistic. If we see our own shortcomings, and say to ourselves that we can do our best to correct them, that is optimism. We have no right to lose our faith. Humanity will not be wiped off this planet. There have been many cataclysms in the past, and we survived them.

We must not be a part of the wrong of the world. On the battlefield, we must live up to the FourSquare Principle. We must prepare ourselves to win the battle.

Cooperation is needed everywhere. Scientists are beginning to recognize this basic principle. Everything is done according to a definite law.

A child wants to be born because that is the aim of the Soul which incarnates. There must be cooperation between the child being born and the mother. The child could cause the death of the mother.

A person living alone in the country, far from their people, cannot get along without cooperation. When he plants seeds for his crops, he needs the cooperation of the seed, the soil, the rain, the sun.

Every element in Nature, the gases, liquids, minerals are all working to cooperate with each other.

When we are in tune with Nature, we are in tune with the Infinite and with our Higher Self. The lower self tries to upset us. The Higher Self tries to cooperate and to coordinate.

We all have, in our own way, to fight something. We must have cooperation.

We all must fight boredom. This is very important.

At the end of our life, the worst fight is the one with death. For most people death does not come as the gentle angel. There is fight, fight, fight from beginning to end. We should all have a soft ending as well as a soft beginning.

We should not say, "I am fighting." Instead we should say, "The Eternal is fighting with me. I cannot do anything by myself, but with the help of the Greatest Power I can do everything I am supposed to do."

We should accept human cooperation when offered, but we must not rely on others. Cooperation not given with the spirit of happiness is useless. Therefore cooperation must be voluntary.

Cooperation means sharing. Why can we not understand this on the biggest plane? Why do we not want the cooperation of the Greatest Power?

In view of our present condition, where we are all, individually and collectively facing the greatest problems we have ever faced, we should realize that "By myself I cannot do anything, but I have a trust in that Power which can help me." We can each bring this realization out from within us.

If we are suddenly reminded to close a door, that is our Higher Self and not our subconsciousness. Our subconsciousness can be trained to remind us, but it works automatically. When unexpected things happen, only our Higher Self can help us. When something tells us to remember something, it is our Higher Self. It is much closer to us, working for us, than our lower self. If we realized this we would be much more optimistic. Just listen to our Higher Self and do as It wants us to do.

There should be cooperation between our Higher Self and our conscious actions. Our superconsciousness trains us and we train our subconsciousness. We are most reluctant to have our consciousness trained by our superconsciousness, but we welcome with open arms all suggestions from our subconsciousness. Why is the hissing and jabbering of our subconsciousness so much recognized? Because we are so conceited; this in addition to Polarity, and our training and experience.

When we are not aware of our Higher Self, we often do as It says. When we become aware of It, then we are not willing to be trained by It. We should welcome a teacher which comes from ourself. Such a teacher is continually at our side. Our Higher Self, working through intuition, should be given more prominence than any other form of teaching. This is not done because subconsciousness is so self-centered. Unless subconsciousness becomes like a little child, it will never learn its lesson. We will never enter into our Higher Selves.

Psychologists not only neglect, but deny the Superconsciousness. They even confuse it with the subconsciousness.

Our Superconsciousness is helping us more than anything else to work out the Principle of our FourSquare.

Memory needs to have a button pressed. Superconsciousness presses the button.

When we deliver a speech, the array of words comes from our Superconsciousness if it is a living, inspired speech. I use this method in my teaching. No matter how sick I have been some nights, I just contact the Power and the unprepared lesson unfolds from my Higher Self. After the lesson I cannot remember anything I've said. This is proof how well Superconsciousness can function. It is proof of the value of Superconsciousness. This is why we should trust it and develop the use of it. It will never leave us without the needed help.

When we make a mistake, Superconsciousness always corrects us. Nothing is too big or too small for it. There is never any difficulty for it in facing any problem.

Superconsciousness says, Dare and Do. It is the adventurous life. Life is a complete adventure from beginning to end. Adventure is the spirit of Eternal Youth. The more we cultivate the spirit, the younger we become mentally and to some extent physically. The lower self has old age and death in it. The Higher Self has eternal youth and life without end in it. Listen to its voice.

The Higher Self is not only represented in humans, but also in animals and plants. The triune being is in all creation. When a dog awakens his master in time of danger, it is the dog's Higher Self. There is brotherhood in all of Creation, and this is contacted by the dog.

In business, that which is called hunch, and which is so important, is the Higher Self.

If we would follow more our Higher Self, we would by and by not only make no mistakes, but we would unfold our FourSquare. The FourSquare comes from our Higher Self. We have always had it, but we must realize it.

Each time we use the Higher Self, we become more encouraged that we can use it. It always gives us guidance and protection. This is wonderful. Spiritualists have a guide in the Beyond. Why should we depend on such a guide when we have our own Higher Self? Who is better off, the person who holds on to his mother's apron strings, or the one who stands by himself? We should rely on our Higher Self rather than on human opinion. Often these agree. This is so much the better. Superconsciousness never doubts. Doubters never use Superconsciousness.

Our Higher Self never says please do something. In the World of Perfection, there is no begging. A general in the army gives orders. He never says please. It is the same with our Higher Self. People who speak with authority speak from their Higher Selves. People who have no authority in them are nothing.

The more we listen to our Higher Self, the more we are creators. Composers of music listen to their Higher Selves. Everything beautiful coming from humans comes from their Higher Selves. We should be guided by our Higher Self and produce more and more beautiful things.

Since we have the three-fold division of mind, we must, in order to achieve Harmony, arrange them so that none of the three shall predominate. The Higher Self should be dominant, in that It is the Leader. Why is it that the Higher Self can be the leader when the Conscious Self seems to be the leader? There is a saying that that which is on the throne is not the most important; that the power behind the throne is more important.

The Conscious Self knows but little of itself, but believes that it knows all and can achieve through reasoning all it wants. If we start our reasoning from a wrong premise, we reach a wrong conclusion. Only if our reasoning is inspired, can we calculate more or less correctly the result. Our reasoning is responsible for all our scientific discoveries. These have not brought us Harmony or a satisfactory civilization.

People judge their civilization by their mechanical developments. We cannot judge by such means. Civilization is supposed to refine people. The horrible weapons of war are not products of civilization. The mind has been used in the most destructive programs man has ever known.

We find lack of common sense in every direction. Something is entirely wrong with our Conscious Self. All mechanical problems are only a small part of civilization. This is the problem facing Humanity.

True civilization cannot perish. It cannot be destroyed. Civilization is first of all a mental thing and not a physical one. If we do not know this, it shows that we are barbarians. Our Conscious Self has gone completely haywire.

The Higher Self has wisdom, but no conscience. Only the Conscious Self has conscience. This is why some religions say that certain people have a burned up conscience or no conscience. They are governed by their subconsciousness, which also has no conscience. The Higher Self needs no conscience. Only the Conscious Self needs it.

The recognition of a mistake is already gaining the upper hand of the mistake. There is no enjoyment of life if something destructive enters in. If one ruins his health in accomplishing something, this is not enjoying life. Life should be at its best when people have developed the finest in their characters.

The mind makes the body age. The body has no power over the mind. We should fight intelligently everything destructive and not submit to it. We permit our air, water and soil to become polluted, rather than to purify them. This is not civilization. We are using our Conscious Self disharmoniously.

The problem of an Utopia is in each human being. The individual's character is only to be measured by the genuineness of his character. If we live Science Of Being, it takes care of our daily life.

Our human mind always tries to make an alibi. The reason people did not foresee our present condition is because they used all sorts of excuses and alibis. People do not stick to the proper standard of "What's wrong is wrong and what's right is right." We have the power of discrimination to do this.

Who helps us to do the things of our Conscious Self? If we would rely less on others and more on our Higher Self, we would be better off. Birds and animals do not need maps for their travels, as we need road maps. We do not want to be helped by our Higher Self.

The life of the whole of Humanity, as well as our own life individually, is at stake now. It all depends on how the last hand is played. Victory can be won at the twelfth hour. Unfortunately, Humanity will not do this.

Superstition is a form of fear coming from the subconsciousness. Never before has there been a greater quest for the supernatural. This is always true when an empire is on the point of declining. This time it concerns the whole of Humanity and not just one country. It is a major condition of the whole world today.

We cannot stop the world sliding down now, but we should learn to slide down intelligently. We should try not to lose our heads. There is not much to cheer us up now. The only thing that can give us some kind of peace and help is within our own self.

Let us try to be friendly, no matter how bad things are. It is very difficult. Be friendly to our own self, to our fine qualities. These are our true friends. Our undesirable traits may desert us when they see they are losing out. Our fine qualities will never desert us.

Let us try to realize that the Eternal is surrounding us in Its four dimensions of Life, Mind, Truth and Love. This is very practical.

To bring out and rely on that which is the best in us is one of our prime problems, because reason and experience have proven that these qualities are the saving elements in our life. We find these in our Superconsciousness. Qualities in our consciousness are of value only if supported by our Higher Self.

It is interesting now to make an historical review of how life has been unfolding on this planet. The achievements of those who have been outstanding have been only of the material kind. Are they of real value? No, by no means. The achievements of the conquerors of the world, such as Alexander the Great who at the age of thirty-three had conquered more of the world than anyone else of that period, what of them? Nothing is left now of those achievements. What is left of Babylon with its great material achievements? Just a desert and memories of something that was supposed to be great. All those rulers who put so much emphasis on material achievements, some are not even known now by name, their palaces and temples are ruins visited by tourists. What are the achievements of various countries of Europe? They built beautiful palaces and castles, and the people lived in them in a very unsatisfactory way. Now their buildings are museums. Will it help people to wander through these buildings? I wonder. The final way the buildings are being used is not the way they were planned to be used.

The Mayans were so developed scientifically that they had a calendar that is better than our calendar. What is left of their civilization? Ruins in the jungle. Where are the Mayans? No one knows.

In the city of Angkor Thom in Indo China, the palace known as Angkor Vat was one of the most beautiful buildings of all time. It was so well built that it withstood earthquakes and other forms of destruction. It is now standing in a jungle. The people are all gone.

All of this is a lesson for us because we are not living here permanently. There is nothing material that is permanent. The people who built all those beautiful things were not advanced spiritually.

In recent times, the cities of Europe were very beautiful, each in its own way. What is left of them after the Second World War? Nothing but ruins. The people have no money, no time and no desire to rebuild them.

People begin, by and by, to lose faith in material things. This is spreading throughout the world. There is no inspiration or protection in material things.

In the United States we have achieved the greatest material unfoldment of all times. We have people who have achieved seemingly wonderful success. But we are not more happy. We do not enjoy the recreation we have here.

We find an increasing dissatisfaction in the whole world and especially in the United States. We are slaves to money, which is a more terrible master than the kings which the former slaves had. It is more easy to get money than to use it properly. Money, in our human concept, represents the Energy of the Universe. It should be considered as something sacred. It is not.

The motivating power back of acquiring money is one of our greatest shortcomings. It is greed. It never pays to be greedy. We should take advantage of an opportunity for the advantage of other people, but not for greed.

Financial and other conditions in the United States are more and more upset because of greed. The United States is perishing now from the cancer of greed. Cancer is a malignant growth of unhealthy cells on a healthy body. This defines greed very clearly.

Humans have cancer in various parts of the body. Communism is a mental cancer. Karl Marx, who spread the cancer of Communism, was later on affected by cancer of the brain and died from it after unbelievable suffering. Wherever he is now his mind is going through unbelievable torture.

What is back of a cancer? Greed. The reason we have so much cancer in the United States is because back of it is greed of some kind. Greed in a certain particular line causes cancer in a corresponding part of the body. The saying, Mind over Matter, is absolutely correct.

Why do we have so much heart trouble today? It is due to our emotions. The heart stands for Love. There must be a complete upset in our functioning of love to cause so much heart trouble.

If a great cataclysm comes, it may bring about a better balance between the material, mental and spiritual in Humanity of today.

People today have turned the Pyramid of Life upside down on a large scale. The base of the pyramid is the material. Then comes mind, with the Spiritual at the top. Today the pyramid is standing on its point. If people did not feel the instability, they would not be looking so much for security. Spies are selling their country for money, for "thirty pieces of silver."

We have so disregarded the Laws of Nature that we call Nature our servant, when She is really our Best Friend.

We are physically, mentally and emotionally cells of our country and of the world. Whatever affects one, affects the whole. The world is going down and we can do nothing about it. We hear of people all over the world who feel that something terrible is going to happen.

If the world is going to collapse, we should try to not let it disintegrate us. Each one of us has fine qualities. Give them an opportunity to express. We should say, "I will find within me the Force to weather the trouble." Fight with the right weapons, not weapons to destroy someone else. With the help of our Higher Self we are bound to succeed. The Higher Self is a channel for the Great Law to express through. If we use it, we can expect to come out with as little harm as possible.

The pyramid is more than a symbol. It is a proof of how a material unit can give us proof of a spiritual one. The pyramids of Egypt were built from the bottom up, step by step. When the top was reached, the pyramid was a series of steps, up which one could walk. Then the steps were covered by a stone which made the sides of the pyramid smooth. This work was begun at the top. The smoothing down was from the point to the bottom.

Our Pyramid of Life started almost immediately after we appeared on this planet. Our progress has been due entirely to inspiration coming from our Higher Self. The urge of living is due entirely to the Higher Self. Without the Higher Self we would be in a state of suspended animation. The Higher Self is a Light at the top of the pyramid, shining down and benefitting that through which it goes. It is more than a light. It is that which is smoothing down, from the top, our pyramid of life. We are smoothing it now. There is not much time left.

Humans think there are short cuts. We must go step by step. No step can be missed. Some people can do it faster, but they cannot skip a step. Quickness is inspired from the Higher Self. Hurry is motivated by fear and comes from the lower self. We are living today in hurrying world. We make a labyrinth of details and lose the big things. Life is going in a zigzag furrow and not in a straight one. Conflicting interests are due to ignorance and lack of being guided by our Higher Self.

In the beginning on this planet there was no consciousness – only the Higher Self and the lower self. The first idea of striking flints to start a fire was due to help from their Higher Selves. They did not get the idea from their consciousness. At the present time we are vacillating between our Higher Self and lower self.

In building our pyramid, there is one point to be careful of. That is: don't wish, but do. Otherwise we will be living in a fool's paradise. To be a jack of all trades is not a compliment. An individual can be richly endowed with talents, but there must be one talent that is dominant. The others are secondary.

The mind is fundamentally a liar. When it becomes honest, it is guided by the Higher Self. The Higher Self must become dominant in us.

The subconsciousness is a "breaker into details." The Superconsciousness is a builder. The scientists get their atomic energy by breaking up material substances when there is plenty of free energy that could be contacted. They get only about one-millionth part, which is only like perspiration. This is proof that Humanity is dominated by the lower self. The lower self is always wrong. It is a negation of the Right.

The subconsciousness is such a tremendous problem that humans will still be solving it for thousands and thousands of years to come.

Our Higher Self directed our subconsciousness to build our bodies. That is why we have such wonderfully designed bodies. "I Am The Ray Of Invisible Light Shining Even In Utter Darkness."

Everything in our subconsciousness is a slave. The Superconsciousness is a free unit. Slaves want to be lazy. The great Law makes subconsciousness work. Slaves should be well fed and treated justly, but not treated in a soft way. They must feel the iron hand. Our subconsciousness should feel this all the time. Science of Being teaches us to learn to do it. We should never be influenced by either our own lower self or that of others.

You may ask, "Is the treating of our subconsciousness as a slave an attitude to be really considered? Why should we in a free world want to enslave anything?" It seems to be a paradox. The heat of volcanoes has at times been enslaved to be used for utilitarian purposes. Subconsciousness is a volcano within us. There is no reason why it should not be enslaved.

In the United States, freedom is the proper use of the wonderful heritage we have, to be free to do the right things and not to do the wrong things. Everything on this planet is limited. So should our concept of freedom be limited. We should take the same attitude towards our subconsciousness. It is a Frankenstein which is enslaving us more and more. This must be curtailed.

Today our subconsciousness has a great deal in it that is useful to us because it has been helped by our Higher Self. We should keep subconsciousness enslaved until it is so reduced that it can no more do any mischief on this planet. At that time there will be no more subconsciousness left. We are educating subconsciousness from slavery to freedom. A problem child must be educated to do the right things. So with subconsciousness, we should deal sternly with it, and with no excuses or alibis.

In the United States, young people are not properly brought up and so are more and more enslaved by their subconsciousness. This is why the United States, and we individual, are in the greatest test we have ever gone through. It is again stated that the Lightbearers are not contradicting the right to freedom.

The Higher Self is identical with our Soul, which is, in reality, our Being. The lower self tried to steal part of the territory of the Higher Self. It took 49% of it. If it had taken 50% of, there would have been a continual conflict between Good and Evil.

Some religions do claim that Hell is eternal. If the Eternal had forever a sore spot in Itself, then the Eternal would not be omnipotent. Because we have 51% we are safe. To those who have, more will be added. Superconsciousness had 51%, and so more will be added all the time. In war, sometimes one side wins for a time and sometimes the other. In the end the constructive side wins and the destructive side loses.

The Higher Self is a magnetic needle and is not affected by conditions. It is the Ray of Invisible Light shining in utter darkness.

Lesson Six

LAWS

All life on this planet runs on a schedule. Planning is so continuous in our lives that we can take no steps without it. Much planning is unconscious. All planning of this kind belongs to the Higher Self. A good dancer must attribute success to his Higher Self. The Higher Self is the Great Architect for each of us. Those hurt in accidents usually did not plan properly.

Back of planning are all the Laws taught in Science of Being. No matter what we do, we must use a Law of some kind. When a dress is cut to a pattern, the law of Analogy is used. If the dress is beautifully made, the Law of Harmony is used.

Refinement is spiritualization. Every disharmony is vulgar. When the lower self influences people, they are vulgar. Vulgarity is in the subconsciousness. Diseases are vulgar. This is why a normal person hates hospitals. Some doctors and nurses face vulgarity to overcome vulgarity. This is very noble, but how many are thus motivated? People with vulgar bodies sometimes have a refined mind. This is due to the Law of Polarity.

All scientists agree there are some unvarying, immutable laws which govern the Universe. It is like a tree with its trunk, branches, leaves, etc. Some of the branches, representing laws, are known to scientists. The leaves are divisions which scientists discover and use to improve life on earth. All leaves will never be discovered. Eventually, when we cease to be human beings, we will gravitate in that direction. This will help us wherever we will be at that time. This discovery of the unmanifested form will stimulate us more and more.

We will proceed, in the explanation of Laws, in a systematic and logical way, not by induction, but by deduction from the Great Law.

Lesson Seven

THE GREAT LAW OR LAW OF HARMONY

What is the most important thing for humans on earth? Harmony. It is also essential to animals, plants and electrons. This Law should always be the prime interest and concern in our lives. Thus we can more and more liberate ourselves from the unpleasant thing on earth.

We should try to find the cause of disharmony. We cannot find harmony in any other way. When something goes wrong, that is a proof of disharmony.

Harmony is so embracing that none of us can approach to the realization of it. We can and must improve. People would never worry how a thing will come out if they did the right thing.

Harmony should be exercised in a two-fold way. We should bear in mind the saying, "Do unto others as we would have them do to us." We should consider both ourselves and others. People think only of themselves and not of others.

Religions, instead of helping people, have handicapped them, because the original teachings were misunderstood. Religious fanatics do anything to win their own salvation, even undergoing physical, mental and emotional torture. This is not Harmony.

No one knows what Harmony is. The best example is found in sound. The more harmonious the sounds are, the more harmonious will be the effect of the sound. At present we have reached the limit of disharmony in the sound of atomic explosion.

Light is Harmony. The sun is very beneficial to most plants.

The sense of smell is Harmony. Fragrant air stimulates us.

Touch. Everything harmonious that we touch benefits us. When we shake hands with people, we should try to register what we feel.

Intuition combines in it the five senses. We see, taste, smell, etc., intuitively. Intuition is the coordinator. We must develop it. By contacting Universal Energy we open the door 90% to Intuition. We should not wait for big occasions to do this. We should train ourselves with little things.

Harmony is essential in every direction. The five senses help us to achieve it. The Sixth Sense is the coordinator which never hesitates and never doubts. In each of the five senses is implanted doubt by the subconsciousness. Doubting Thomases let their subconsciousness rule them.

We have to learn to know what authority means. It is the operation of the Law. It is based on the Law of Harmony. We should develop the sense of authority. Most people use it in the wrong direction. Right authority always produces the right effect. To develop authority in the right direction means to develop our sense of Harmony and to be the Image and Likeness of the Eternal. Jesus spoke with authority. I notice that we are all nearly completely lacking in this. Mistakes show the imperativeness of authority.

Harmony is not only the way we feel inside, but the way we act. We should always remember there is a Higher Power that can help us. "In Thy Hands do I place myself."

The Law of Harmony has been perceived by people since their very beginning on earth. They have developed their innate sense of Harmony.

The FourSquare is like a magic sound to THE LIGHTBEARERS. It is the symbol of the Eternal, the most sacred thing in the Universe. It is so sacred that when ignorant people use the word FourSquare, or square, they usually bow their heads. A luminosity often appears in the faces of the worst individuals. It is the Father and Mother.

The triangle of Mind in the FourSquare is the triangle of the Father. The triangle of Love is the triangle of the Mother. All triangles meet at the point, all touching each other.

The FourSquare is unbreakable, therefore eternal. We are supposed to be the image and likeness of the FourSquare. The Eternal contains us. We are a little FourSquare in the Eternal FourSquare. If we manifest as much as we can our FourSquare, we are in touch with the Eternal. When we develop the FourSquare within us, Harmony is there.

Our Commandment has Harmony in it. It is the expression in words of the FourSquare. We must use it wisely and wholeheartedly. Use Love as much as possible.

The Great Law and the FourSquare are identical. The Great Law manifests through countless smaller Laws. The FourSquare is a symbol and also a key to understand the Great Law. The Great Law cannot be used in a practical way because it is not understood by mortals. The FourSquare makes it simple, understandable and practical. If human beings since their beginning on earth had carried out the principles of the FourSquare, life would be not only interesting but most pleasant.

In the world as it is today we see an ever increasing unbalanced condition. In everything and everybody, one point of the FourSquare is developed without the other points. This so confuses the people, they don't know which way to turn. Disharmony is a direct product of an unbalanced condition.

We have energy. Use it properly.

We have intelligence. Use it properly. We are conscious of our own intelligence, which makes us different from the rest of Creation. Intelligence should never be influenced by blind understanding.

We have truth. Use it properly. Everybody has an inborn urge to be truthful. The corner of Truth can be traced to the very beginning of our coming to earth. We came as a result of the misuse of Truth.

We have love. Use it properly. Everything is born with a spark of Love in it. Nature uses it for the continuation of the species. Love is the keystone of all existence.

"With the help of the Great Law" means "With the help of the FourSquare within us."

Lesson Eight

LAW OF ANALOGY

There are no thousands of Laws governing things separately. There is one Law, and that is the Law of Justice. The saying that all men are born equal is not true, unless we add to the saying the words, "before the Law." The Law takes care of each, in Its individual way, not in a standard way.

We should try to understand the application of the Law of Analogy. If we do not know how to handle an emotional trouble, we should translate it to a purely physical aspect. If we find that it will work in a certain way physically, it will do the same emotionally. Do the same thing with a mental trouble.

Each plane has divisions. We can judge big things by little ones, and little ones by big ones.

Scientists could use this Law greatly, but they do not. They are affected by doubts and distrusts in their own mind. They think the Law of Nature does not always work in the same way. It may seem to work differently because the channel is different.

If we do not know how to do something, try to find an example which we do understand and then apply the Law of Analogy. As an example, we know that seeds should be planted in order to make them increase. This analogy may be applied to each of the four corners of the FourSquare, as follows:

ENERGY: Money should be planted. The United States Government, instead of planting and cultivating its gold, buried it at Fort Knox. This made out of the richest country in the world the poorest. We are now absolutely bankrupt, although the people do not realize it. Richness is that which is flowing over. The United States Government has to squeeze out of the people, by taxation, the last drop. This is poverty.

MIND: We must plant wisely. Professor Felix Ehrenhardt, directed probably by the Great Law, came to the United States and made one of the greatest discoveries of all time, that of a machine to contact Universal Energy. The United States would not accept it because he did not present to them from the view point of how they could make money from it. Instead, he approached the whole matter from an altruistic view point. It was not accepted, and he returned to his native country of Austria, where the Communists probably will make great use of it. He did not plant wisely in this money-conscious United States.

TRUTH: Every new discovery is a seed of Truth. It will only grow when the soil is ready for it. Atomic energy was a great discovery, but it was planted in distorted minds, who were not ready to use it in a beneficial way. The teaching "Love thy neighbor as thyself" was planted in minds that were not ready. As a result there are hundreds of thousands of churches with the wrong understanding of these fundamental principles. There is now destruction all around us instead of Love. The church buildings are sepulchers with beautiful monuments. The United Nations built a very big building in New York in the shape of a huge tombstone. It will be a memorial to show where the United Nations is buried.

LOVE: The seed of love planted in the wrong soil will give disappointments and hatreds. The relation between husband and wife is supposed to combine the physical, mental and emotional. There are so many separations because the seed of love was not planted in the right soil.

The basis of good business relationships is love. An employee should feel that he is a part of the business. If the business is benefitted, the employee will be benefitted. Employers should not consider an employee as a mechanical unit. The employer should be a father and a good shepherd to the employees who are his children and sheep. The employer should be interested in the people who make the business a success. The employees should be interested in the business. It is all based on love. To do the right thing is to live up to the corner of Truth, and this is the gateway to Love.

People should be good neighbors, but we should choose those with whom we are intimate. There is a way to be friendly, but not chummy.

Lesson Nine

LAW OF VIBRATIONS

Vibrations may be divided into three groups – life or physical, mental, spiritual.

LIFE VIBRATIONS: Life vibrations are the essence of our lives. The more we understand them, the more we understand life. The more we understand the demands of life, the better can we shape our own life.

Where there is life, there is Law. The two are inseparable. In the present condition of the world, one usually predominates over the other. Life has become, instead of a boon, a trouble. Instead of Harmony, there is disharmony. There are those now who have life without law; and there are those who have law without much life. Both are wrong.

Those who are so-called successful live a life without Law. They are guided by ambition, greed, lust or sometimes by their hatreds. They have plenty of life vibrations, or energy, and therefore they succeed in the eyes of the world. They are often dead, or insensitive, to human laws, and therefore escape punishment. Those who are supposed to apply the laws to them, do not do so. In time, however, the Law of Retribution acts. The ultimate verdict is that crime does not pay.

There are no end of little crimes committed all the time by those who try to live the Law. These people have to pay for them.

Trouble extends itself over those who are close to the ones who break the law. It often seems harder for relatives, or close friends, than for the law breakers.

When one lives only the vibrations of life, these vibrations can be very destructive. This is why it is very important that we understand the importance of life vibrations. The Law of Life Vibrations is a fundamental Law. It gives power to the vibrations of mind and of spirit. All are fundamentally connected.

Life vibrations should be the first subject of our intelligent approach. Instead, people seem to disregard these more than anything else. Those who are so much concerned about their health really only think they are. They eat certain foods or do certain exercises because it is a fad to do so. Life is not a fad. It is the most serious thing on earth. The blame should usually be placed on the doctors, who are supposed to know more than other people. These disagree among themselves to such an extent that the people are confused.

Animals live a much more intelligent life, because something within them guides them in their proper food, exercises, etc.

It is not easy to use intelligence in such a confusion. We should rely more on our Higher Self than on our reason.

Lack of consideration and lack of compassion is very prominent today. We should realize that somebody else has as much right to live as we do. Consideration should never be exaggerated, but should be given when, where and to whom it is needed. We should never damage another. In home relations, consideration is not shown in most families today.

There should be respect for old age. If it is not given, it is like stabbing a person in the heart. This is not living Love. This condition is very strong in the United States today. When a nation begins to lose respect for the accumulated wisdom of age, it is a definite sign of decay.

Killing of time is not respect for life vibrations. Life should be lived.

In the present world, people seem to be determined to kill as many others as possible in wars. This is a perversion of life vibrations. The destructive spirit of children who break electric light bulbs on streets is due to unhappy life conditions. We deteriorate life vibrations all the time. Life vibrations are very precious.

We should do things in a harmonious, gentle way. Life vibrations should be intelligently used. Think first and act next. Some people are paid for wrong thinking, others for not thinking at all. Is it worth the trouble to live such a life?

Lesson Ten

MENTAL VIBRATIONS

All vibrations are based on energy. Energy is One, and in the Realm of Harmony, there is no division, but there are rates of vibration.

In the rainbow, the rays of color melt into each other. There is no sharp division between them. The rainbow is symbolic. It describes vibrations in the Realm of Harmony.

In the world today, mental schools recognize only mental vibrations, while spiritually minded schools recognize only spiritual vibrations. These are both wrong. Their beliefs are not according to the Law of Nature, as illustrated by the rays of light in the rainbow.

Our mind is a great gift of the Eternal to us. It is the projection of the Universal Mind into Its own substance, indissolubly connected with the Eternal Intelligence. The Eternal Mind is incorporated into our own mind. We do not understand one one-thousandth part of the power of our mind, because it is a divided mind, fighting between the lower and higher divisions. The battle is taking place in our conscious self.

We should understand one fundamental principle. It is the Principle of the FourSquare. It is always the foundation and the ultimate of our activity.

1. Put all energy into our thoughts. No progress is possible without concentration. We are living in a fog. The more we can clear our mental vibrations, the better off we will be. To develop concentration, first be attentive. Be earnest in all we do. Earnest means concentration.

2. Think as wisely as we can. Judge things as correctly as we can. If in anger, don't say anything. If motivated by greed, passion, etc., think the whole matter over first. See if it is the right direction. When certain, act instantaneously.

3. Be sincere. No one can do anything without being sincere. Sincerity pushes a thing forth. Sincerity is an aspect of Truth. No human can be sincere without being truthful. One of the worst human shortcomings is to be a Pollyanna. A Pollyanna tries to make wrong appear right. They cannot be sincere. Leave our enemies alone. Do not cast pearls before swine. I do not think that Jesus said to turn the left cheek, but rather to turn away the left cheek.

4. Be joyful in our thinking. Do things joyfully. Hope is the greatest element in doing things joyfully.

   If we use the FourSquare in this way in our thinking, it will be of great advantage to us and to all in reach of our thinking.

Telepathic transmission of thoughts is now a proven fact. It is known that if a very important idea is brought into manifestation by a strong individual, it by and by spreads all over the world. Those in tune with it are the first to receive it. Those who are indifferent to it, receive it last. Those who are opposed to it, fight it. These, when the world accepts it, are the first to claim that they accepted it. This invariably happens.

People now feel that world conditions are getting from bad to worse. It is like a tension reaching its limit.

There is now a mental breakdown. The individual's mental tension increases to such an extent that the individual's nerves can no longer throw it off. Then there is an explosion which is called a mental breakdown.

This situation is now in the mind of Humanity. It may have started centuries or even thousands of years ago. There have been, in the past, small alleviations, or safety valves, but these safety valves have not increased as fast as the tension. Safety valves are the art of relaxation.

To sleep was one of the relaxations. Another was a change of activities, such as dancing, singing, playing music, various sports, etc. The main relaxation was joy. Each person has his own approach to joy.

Today real joy has been put more and more away, and replaced by excitement or thrill. This is so because we are in too much a hurry. Joy cannot be hurried. We now entertain and are entertained, but we do not enjoy it. This, by and by, makes us forget what real joy is. To say that it is a nice day is not enough. We should actually feel the nice air, the sunshine, etc., because Nature is very bountiful in every way.

In days gone by people enjoyed drinking by preparing themselves for it. They would first smell the drink, then sip it very slowly. This is real enjoyment in drinking. Today they gulp it down, and then wait for the after effect.

Unless people were very hungry, they would enjoy the odor and flavor of various foods and then eat gradually more and more of it. They had better digestion.

Today people gulp the food down, not getting real enjoyment from it, and they ruin their digestion.

People in older days prepared themselves for every activity. Now they lose more and more the sense of true enjoyment.

Religions took more and more real enjoyment out of life than anything else, and substituted for it an imitation of enjoyment. Instead of spontaneous singing, they organized it, in dusty churches, often by singers who did not know the ABC of singing. People have revolted against religions because the tension against them could no longer be endured. There is now a lack of true enjoyment which people got from primitive religions and from Nature. There are now restrictions on everything.

A thrill can never produce lasting results. We are always in search of something, by thrills, which always evades us. In the olden days people not only got real enjoyment, but they retained it. The art of real entertainment is one of the greatest arts to be cultivated – such as by music, arts, etc., and rarely by silence. There is true peace, true joy, true simplicity which people do not understand nowadays.

What has all this to do with mental vibrations? What we think, we are. True harmony, true civilization lies in our mind. Every organ in our body would remain harmonious, if it were not for the interference of our mind.

There is a thinking now all over the world of a feeling of anxiety. People feel that something physical is going to happen, such as an atomic war, caused by humans, or an earthquake caused by Nature. The Great Mother Nature has never been allied against us, but our mind has perverted the Laws of Nature. Before any such catastrophe can happen, there is a mental tension and a physical tension. We now have reached an unprecedented condition of this kind. We cannot stand it much longer. The more primitive people feel it will be an earthquake because they are more in tune with Nature. The more sophisticated people think it will be an atomic war. Which will it be? The more we feel a thing, the more we think about it; it will by and by become a reality.

What precautions can we take? The precaution of our own mind. The best protection is a harmoniously functioning mind. Do not lose ourselves, or get panicky. Do the right thing at the right time. This protection can only be achieved if we establish in our minds a sense of harmony, peace and joy.

The greatest termite in our minds is resentment. It undermines the whole of our mental peace and harmony. Resentment is a cancer of our human minds.

Hatred burns itself out. Resentment never burns out, but burns deeper and deeper and deeper. I do not think that Jesus said to turn the left cheek, but rather to not be resentful if hit on the right cheek.

The world is now full of resentment. This brings out and nurses all the various secondary traits in our subconsciousness. This storm which is now menacing Humanity can only be faced if we are solid mentally. If the mind is not solid, the body cannot win out. The whole chemistry of the body is influenced by the mind.

Resentment is the wedge through which everything negative can establish itself. If we could eliminate resentment, and we must some day, we would feel much freer. Resentful people radiate resentment and awaken resentment in all people with whom they come in contact. Resentment is so subtle, because we do not see it, as we see hatred in a person. We can feel it, and when we feel it we know a person is radiating it. If we show to such a person that we are not resentful, that person becomes more resentful. We must keep up vibrations of Love until the resentfulness in the person completely dies out. We will be mentally destroyed by the world cataclysm if we do not overcome resentment.

We should be wholehearted mentally. If we hate, or if we are greedy, we should acknowledge it. Let us try to forget the wrong which people do us and try to think the reverse – peace and joy and harmony.

Lesson Eleven

SPIRITUAL VIBRATIONS

It is not easy to discuss Spiritual Vibrations, because the mere uttering of words and phrases is not enough. This is enough for physical and mental vibrations, but Spiritual Vibrations demand our complete surrender to what the vibrations mean and to the essence of the vibrations. This is very difficult in the present condition of the world.

What are spiritual vibrations? To analyze them, we must try to explain with words what we must sense when we merge into the Realm of Harmony. First, forget oneself and say, "I can do nothing by myself, but with the help of the Father (or Great Law) I can."

The Spiritual World is governed by an Eternal Law. This invisible Eternal Power is back of all physical expressions.

The North Pole and South Pole are symbolic of cold. There is very little animal life there. Vegetation does not exist at all. Life is almost circumscribed to the mineral. The Equator represents life in its most glorious unfoldment. This is due to the sun, both the physical and the Eternal Sun. Without attraction, the Universe could not exist, function or remain in a state of stability. The more there is Love, the more there is life.

The long nights and long summers of the Poles represent a cold balance. The balance on the Equator is more equitable, more joyful, more warm. It enables us to live a much more happy life on earth. It gives us a sense of inspiration and beauty beyond description. There is very little of this at the poles. The mountains of ice are beautiful, and also the Aurora Borealis, but it is all a cold beauty.

Mountains at a distance inspire us, but when we climb to the snow-covered tops, it is beautiful, but we feel solitude, the most difficult thing on earth to bear, unless we have risen above it.

It is the same in our present life. In the Spiritual Realm, there must be high peaks reflecting all the glory of the Spiritual Sun, but today they seem very distant. This world is a very cold world. Most of us, by and by, discover how cold it is.

Mentally, we rise into higher and higher strata, but the higher we rise the colder it becomes. As, for example, the cold mental approach of the experiments under Hitler. Human beings ceased to be human. This is spreading throughout the world. Humanity is steadily growing colder and colder, steadily growing away from the highest vibrations.

There are very feeble attempts to counteract this, such as being charitable, helpful, etc., but these are only in self defense, because Humanity realizes in its heart how cold it is. All these efforts are only proof of a growing coldness.

There must be a counterpart of this condition, or one of growing warmth. At present there is a fight between the two. When warmth loses its balance and becomes heat, it is just as bad. It burns. Intense heat is the heat of hatred. When balanced, we have the best condition.

Love was once more prominent on this planet. We can still see it in primitive people. Mind created adverse conditions. The fight turned Love into Hatred.

A mind that wants to warm itself goes to extremes. It hates first itself, then others, until it becomes a consuming fire. We have this conflict today in each of us individually, and in the collective world. The change must start in the individual and not in collectivity.

Spiritual Vibrations can only be found in the FourSquare, where the dominant corner is Love. Any other method, either mental or physical, will fail. We must be an open door to it. That is the problem today. Love is Harmony, and Harmony is the basis of everything worthwhile in life. Everything in Nature senses Harmony. We have drifted so far we often call disharmony Harmony.

We must keep the fire of Faith all the time burning. Use discrimination to learn what to believe and what not to believe.

What and where is Truth? We can learn best by Intuition. Intuition is the key to the Spiritual Realm. No prayers, no sacrifices, nothing that we are taught to believe will ever begin to approach Intuition.

Intuition is based on Love. A mother, even animal mothers, intuitively feels the need of her young, because she loves the child. When people love each other, they know without words what the other needs.

In every direction, Intuition is valuable. It is gentle, but not cold. It has not the official stamp of our human minds. It is above law. Since it is the Higher Law, it can never violate any right human law. Be loving and hold preciously in your hands the key of Intuition. It will fit only in the right keyhole. If we understand this we can help ourselves and others a great deal.

We can never be of help to others if we do not use the principle of help to ourselves.

We cannot get the highest vibrations without faith supporting us. We want to get the finest in our body, mind and emotions. Only Faith enables us to contact the highest, because Faith and Love are inseparable. This is very important to remember. Faith means complete surrender to the Highest. Faith helps in physical and mental contacts, but is not necessary. Faith, like Love, can never be forced.

The present world conditions are far from a spiritual state of consciousness. Probably at no other time in history has humanity been so materially minded. Even so, we are not left entirely hopeless in developing our approach to higher vibrations. Our Higher Self always functions, even if we do not perceive it. It always guides us. It knows what is right. It does not know wrong.

What we call the unfoldment of our Higher Self, is just the perceiving of it by our Conscious Self.

The following are steps which will help in contacting Spiritual Vibrations:

AIR: Think of the air we breathe. We can live without food for quite a while, but not without air. Polluted air is more detrimental than polluted food. Pure oxygen is a symbol of life. Everything in the body functions due to oxygen. Life is the foundation of everything. It is a spiritual quality. By breathing oxygen we take in a spiritual quality. It is the first step towards spirituality.

LIGHT: Everyone appreciates a sunny day. Life lives because of the sun rays. At night we have light from the moon and stars. What is light? It is a spiritual thing. Light, on the mental plane, means knowledge.

BEAUTY: Think of all the beauty Nature has spread around us. Beauty has a spiritual basis back of it. Whenever we see or smell a rose we unfold towards spiritual vibrations. There is also the desire for things beautiful in man-made paintings, sculpture, etc. This indicates the great quest of Humanity for something to uplift them, to reach something higher and better. Try to make as much use as we can of manmade beauty. Whenever we are inspired by beauty, we have taken a step towards Spiritual Vibrations.

When we try to dress ourselves the most beautiful that we can, we are rising towards spiritual vibrations. Animals try to keep themselves clean and to beautify themselves. In the United States especially, humans try to keep their bodies clean.

Body odors are due to food and to the way people think and feel. Cleanliness is next to Godliness, because it is the aspiration of the individual towards something higher. Fragrance of the bodies of some people is due to the high rate of their vibrations. Usually very kind people ooze kindness as a flower oozes fragrance.

LOVE: The feeling of Love implanted in us and in the whole of humanity should inspire us. It is the most precious thing we can have. People who like to be alone have found the partnership of the Eternal. The South Sea Islanders recognized only one sin and that was to hurt another individual. Humanity and religions have not lived up to this. Each time we hurt someone we lower our rate of vibration. If others hurt us, we should say it is too bad they do not know better. Human love is a ladder to step into higher vibrations. Give to the stranger the richness of our smile is a good way to reach to higher vibrations. Let our finer nature inspire us. This is the most pleasant, simple and satisfactory way to reach spiritual vibrations. The good deed per day of Boy Scouts is very commendable. It is an example of many things we do collectively which inspire us. Collective actions of this kind are a proof that humanity is slowly rising higher.

LOOK UP: After realization of these spiritual qualities, or during the realization, look up, with closed eyes. No human who is angry ever looks up. Jesus always looked up when He "prayed." If we look forward, we face life as it is today on this earth, which is a very unsatisfactory life. If we look down, it is worse, because we contact our subconsciousness. The higher we look up, with closed eyes, the nearer we reach the pineal gland, and also the pituitary body. We begin to feel very calm and peaceful. Have faith that we are contacting spiritual vibrations and it will be so. It is a very simple method, but it is not easy to do.

Lesson Twelve

LAW OF POLARITY

The Law of Polarity is fundamentally a spiritual Law. It means, on the Spiritual Plane, harmonious balance and synchronization. In our present state of consciousness, it is incorporated into subconsciousness. This started eons ago, when subconsciousness came into existence. It has now become, together with the Law of Rhythm, laws of the most destructive kind. There is now negation of an Eternal Law. The Law of life is now the law of death; Harmony of disharmony; Knowledge of ignorance; Truth of lies; Love of hatred; Infinity of the Universe of limitations; Analogy of misunderstanding; Nature of vibrations reversed to destructive vibrations; Gender, which was once one, is now division; Cause and Effect, which was a Law of Logic, is now one of confusion.

We must work out of this condition into liberation. This will come only when the last mistake is corrected and the last fault in our character corrected. Nobody knows even approximately how long it will take. Each cycle becomes a shorter cycle. The First Cycle probably lasted for hundreds and hundreds of millions of years. The Second Cycle lasted for an unbelievably long time. The Sixth Cycle, which we are now entering, will be short, but still a long, long time. Then we will step into the Seventh Cycle, when there will be a rest, a fruition of all our efforts of previous cycles.

Today we have more understanding to help us face Polarity than we have had before. Polarity is the most destructive opposition one can imagine. The opposition extends itself through every activity of the human mind. It is the main reason for our struggle for existence. It starts when we are born and extends to the very end of our lives.

We cannot abolish a law, but we can abolish our wrong interpretations of the law. The sense of sight is the sense of perception. All the other senses are supposed to cooperate with it. Our sense of sight was reversed by Subconsciousness, so that we saw a perverted world. Our other four senses did not agree. They refused the leadership of Mind and won the battle. We had to learn to see things straight. The fight probably would not have been won if there had not been still within us the unfallen Spirit of our Higher Self. It is a tremendous Power within us. The four senses cooperated and won the battle.

When we say, "I am well," we should think of the Higher Self, which *is* well. Subconsciousness cannot change the Higher Self.

When we feel opposition, it always comes from our subconsciousness. It is polarity. When we want to do something's right, and something opposes us, look for polarity.

Polarity in its higher aspect is a Law of Harmonization, bringing things together. Polarity and Rhythm, as we see them now, are a wrong interpretation of a right Law.

Whenever there is a mental opposition within us, say to our self, "that is polarity as subconsciousness wants to interpret Polarity, and I am not going to accept that interpretation, but will take the real meaning of synchronization and Harmonization." Thus we will be able, step by step, to change the whole trend of our life. Some day we will succeed in doing better.

Lesson Thirteen

LAW OF RHYTHM

In this world of duality, which can be very appropriately named the world of opposition, we have two laws of opposition working together: The Laws of Polarity and of Rhythm. Vibrations of the same kind blend with each other, characters of the same vibration blend together, and, in the same way, laws of the same nature blend together. This is why Polarity and Rhythm do not work against each other, but instead prevent our unfoldment. Life should be one unbroken rhythm, always rising higher and higher, with no downward movements. But in this little locality which we call the earth, and in those little spots called human beings, the laws work differently.

In a clock, the pendulum is circumscribed by the clock. On the mental plane we are living in our subconsciousness, and the pendulum of the Law of Rhythm functions. Each clock will endure as long as the mechanism will endure. Our mental mechanism has endured for billions of years and is still strong enough to endure for a long, long time. Therefore we cannot expect liberation as soon as many people expect. But mental things do wear out. Subconsciousness once started and therefore it must come to an end. Therefore the constituent parts of subconsciousness must wear out. This includes both Polarity and Rhythm.

If our patience is too much strained, it wears out. If our optimism is too much strained it wears out. Everything, if not replaced, wears out. Our body wears out because the cells are not sufficiently replaced by new ones, due to our mind. Mind wears out quicker than the body. Each cell has its own mind. If left alone it does not wear out, but when affected by the whole mind of our body, it does wear out.

Why do we grow old? Youth is an eternal condition. The Universe is eternally young, and all its manifestations should be the same. Our body is a victim. The culprit is our own mind. The minds of men deteriorate in most cases much sooner and quicker than those of women. This is because love is manifested in men less and less as time goes on. There are exceptions.

The fact that mind wears out seems to be a tragedy, but really is a blessing in disguise. The sooner it wears out the better, but it should do it in a better way. It should resist the process until the last minute of human life. Usually it works this way in animals and plants. They suddenly collapse and die. If we could do the same, we would grow old, but remain fine physically, mentally and emotionally until the end, and then just collapse. I hope that Science of Being, when properly understood by humans, will teach them to keep up until the very end. This is not a dream, but a future – a far away future.

If we understand the Laws, we can integrate to a certain extent. We must be continually expanding, reaching to a greater freedom. If we cannot do this with our body, we must at least do it mentally and emotionally. At present, we deliberately imprison ourselves in our own mind. People who are broadminded have broken away from the limitations of their subconsciousness to a certain extent.

The monotony of life is a tragedy. It takes away the incentive to live. Every moment should be different, something new. Fog is a symbol of our subconsciousness, of the dullness of our life, of the grayness of our life. We can break this tragedy. Then, by and by, the pendulum of life will change.

The best answer we find in our own self. The little sun shining within us is a mighty powerful sun. From somewhere within us we can burst out. Then rhythm is licked.

The voice was given to us not only to express our self, but to free our self. Slaves sang when they worked, because they found relief in it. The peasants pulling the boats on the Volga River sang "The Volga Boatman", a song which I especially like. To listen to music is not enough. We can only be a part of a thing when we perform it. Singing entertains the Inner Self, which in turn entertains us. When the Higher Self expresses through singing, it brings out the very best in us. Thus can we help to lick rhythm. It is so simple most people cannot understand it.

Life must have completion. A flower must be complete to be beautiful. The movement of rhythm is incomplete. It must go continually onward instead of stopping to go backward. We must demand from life the whole. When we do the best we can we have broken the limitation to a certain extent. Next time we can do better. If we have something fine within us, say it or do it. Thus can we break rhythm.

We live now almost at the lowest point of the downward movement of rhythm for the whole of Humanity. When things are in a stormy condition, the test of our character is to demonstrate that we are above water. We are not meant to perish. Life does not want destruction. The mandate of Life is to live.

Lesson Fourteen

LAW OF GENDER

In our present state of consciousness everything has two genders – the male and the female, the positive and the negative. The male and the female appear to us separately in humans, animals, plants, minerals. In order to complete each other, they have to come in direct contact, one with the other. This is for the best, because they have to continue their existence on this planet. Without it there is no possibility for the Law of Reincarnation to work. This division has to be overcome, because it is not the ultimate. Harmony can only be achieved when the positive and the negative come together, each contributing what is necessary to the other.

There are no soul mates on this planet, or anywhere else. This idea was promoted by poets. It would be the worst punishment for any of us to be tied to another for eternity. We may come together in different incarnations, but we are not tied together eternally. We can blend with people, but we should not lose ourselves to them. It is fine to blend as individuals or as a group. Yet, there is a soulmate. Our soulmate is the Eternal. He is our only soulmate. We are never tired of Him, because he always reveals to us more and more of Himself. People brought this idea down to the human level because they could not see big things.

In many cases affinities work better in animals than in humans. They often have the same mate for life.

In the wind, the gentle breeze is the female, the storm is the male, and the quiet is the two in one.

In the ocean, the peaceful waves are the female, the stormy waves the male, and the placid water is the two in one.

In the earth we have the two poles. The South Pole is the male gender. It is protruded outward, as a mountainous continent. The North Pole, which is the female gender, consists of a cavity. The South Pole protuberance would probably fit into the North Pole depression very well. The positive currents of one pole reach the negative currents of the other along the line of the magnetic axis of the earth, yet each pole is predominantly either positive or negative.

The present apparent separation of the sexes is only temporary. The man and the wife should contribute their qualities to each other. They should bring out the static condition of the negative by combining with the dynamic condition of the positive. By continual association both grow into individual, complete beings. The more they succeed, the more they become independent – a complete, rounded-out character. Real Harmony is achieved within each. There are marriages which approach this ideal, but real fulfillment does not exist on this planet.

A great deal of development and good is achieved by sticking together, even when there is not full harmony. Differences can be adjusted, if there is good will on each side – a teamwork with mutual adjustment. It can be done, if both parties try their very best.

The quest for happiness and harmony, and the patience to get it, is one of the most difficult things to achieve. In a world which is in so much of a hurry, it is very difficult.

How can we solve the problem of being patient? Each one must find it in his own way – sometimes by a hunch, or by hard work, or even by failures. Carefulness is a form of patience.

In driving a car, patience or carefulness, which is a trait of the female, combined with speed, which is a trait of the male, gives the best results.

Sometimes these qualities are developed in groups of the same sex. The leader is the male, and the followers the female.

The male and female qualities are both powers, and very beautiful powers. These fine characteristics of a complete being are within us. Let us bring them out. Perseverance, cooperation and collaboration are required.

In the United States we have broken down the barriers between the male and the female. A man who does house work is not longer considered a sissy. This is why we should have full faith in the United States. We are on the right path. It shows the growth of humanity into complete beings.

Lesson Fifteen

THE LAW OF CAUSE AND EFFECT

The reason we are humans is because we forgot about the Law of Cause and Effect before we became humans.

On this physical plane a dream, though it seems very long during the dream, actually is only about a second or two in time. The dream of Humanity through which we are now passing seems to be eons and eons in duration. When we finally awaken in the Realm of Harmony we may find that it was only an instant of Time.

Under the Law of Cause and Effect, we have the possibility, no matter how dark the conditions seem, that if we do the right thing we must and will get the right effect.

Lesson Sixteen

THE LAW OF LOVE

Love is the Law and the Fulfillment of the Law. It must be felt. Thinking about it is not sufficient. The human mind has no power of attraction, except when Love enters. Where there is no sincerity there is no truth.

Achievement is mind and love combined – the Eternal Twins. They are still fighting each other.

The Law of Love is the Law of Universal Harmony, the fulfillment of all we can wish for on this planet, and the only protection we have.

We see in present conditions how the absence of Love has affected all things. Throughout the whole world we hear the call for Love, in the form of Peace. Only Love can give us peace.

On the human plane we have passion as the counterpart of Love. In the United States this side of Love, or passion, which is purely physical, has been overemphasized. On account of this, the United States is collapsing.

True love extends itself not only to those dear to us, but to everyone. We should protect ourselves from enemies by Love. We cannot afford to hate enemies. Hatred is like an avalanche which increases all the time. If people would understand this, they would realize that they cannot afford to hate. The same Law works through the positive and the negative, Love and Hatred.

The world can only be saved by unfoldment within our minds of Love. We are living in a thoughtless age when it concerns Love. The religiously inclined kill another religion. The materially inclined kill another nation.

Love is strong. It is never weak. Are we strong? No. If the desire of people would be sincere and genuine, even our enemies would not attack us.

We see the life of a nation on a large scale. Individuals affect the nation, and then the nation collectively returns the compliment.

Our Teachings are so precious because they enable us to see what is right and what is wrong. Love was always the fundamental teaching of Avatars. There is nothing new about it. Science of Being has Love combined with Wisdom.

Anyone can love those who love them. To have a feeling of compassion towards those who hate us is more difficult. In this age where mind claims supremacy, hatred is very strong.

When someone insults us, if we make a reality of their hatred, we will be insulted. If we rise above it, their desire to hurt us will go back to them and hurt them. We should say, "Poor fellow, he is only hurting himself, he cannot hurt me." We should not accept insult. We should refuse it in our mind. There should be no question of retaliation. This is not religion but is based on Laws of Nature. The only way is to do the right thing.

We are imprisoned in our body, mind and emotions. There are no doors, only windows, in the three-fold prison, but unfortunately the windows have bars, and we cannot break them with human, mental or emotional hands. They can be broken only by steel so strong that nothing can conquer it – "The unconquerable Sword of Truth." Why? Because if we have not learned the lesson to be honest and truthful, we cannot learn the Law of Love. Love has a dissolving power on everything negative, and a power to bring together, hold together and cement together everything positive.

None of us can rise to contact continually the highest Love Vibrations, so we must be contended with Truth. We can compel people to tell the truth and we can force ourselves to be truthful, but we cannot compel people to be loving.

We are bound to our own subconsciousness – the great prison in which we live. We must break through the prison. We must cut the bars of the prison windows. We have only one way to do it. That way is to be honest, sincere and truthful.

Lesson Seventeen

THE LAW OF EVOLUTION

The Law of Evolution is an Eternal Law. It never started and will never come to an end. It is the Law of the Eternal Spiral, going from Infinity up to Infinity.

How is it possible for it to continue from a beginningless beginning to an endless unfoldment? In our subconsciousness, it is not possible to understand it. It is beyond the three-dimensional state of consciousness. It is in that Realm where Freedom is reigning supreme.

We should try to realize that the Eternal has given us the Law to take away the monotony of life to give us in this earthy life more freedom and happiness.

When we have fulfilled our course, after many more incarnations, we will realize what a wonderful Law it is. It works against Polarity and Rhythm. These laws, as accepted in our subconsciousness, are only temporarily working, and though seemingly so strong, they are not as strong as the Law of Evolution, which is a Ray of Light shining in utter darkness.

No matter how much we are swinging from the positive to the negative, there is always a small gain. When we deliberately or otherwise violate all the Laws, The Law of Evolution still works. A wrong cause is bound to culminate in a wrong effect. Evolution works through the penalty, by the suffering we go through. The result is that in spite of the mistake we still evolve. Nothing can stop Evolution. Not even a mistake can stop it. Evolution is carrying us with it all the time. When we fall, we always fall forward. We always cover some ground though sometimes paying a heavy penalty for it. In our present state of consciousness, the Law can be called the Law of Hope. If we understand it we gain a little more courage.

Take present day conditions. What misery is befalling Humanity! Even we Lightbearers cannot say that we are happy. How can we be happy when we see others suffering? All we can say to ourselves is, “Thank God that I am well.” Even in this attitude to the Eternal there is a bitter taste because others are sick. In darkness we cannot realize the glorious sunshine of Life.

Once we were enjoying life such as no mind can conceive. Every moment was unfoldment, each day more unfoldment, and no end to that unfoldment. Instead of being grateful for it, we rose against the All Power Who was giving us that continual unfoldment of life, that Joy of Life. What was the result? One mistake, one wrong cause started, and for countless eons of time, for millions of incarnations, we are still paying, and we will continue to pay. It seems unfair that we should have to endure so much suffering, and for so long a time. We ask, "Will it come to an end?" Yes. Anything that starts must have an end. Only those things that never started will never end. Because it was the greatest offense to the Greatest Being, the penalty, or effect, was according to the mistake.

I feel that the greatest mistake in my life, which is small compared to the colossal mistake we all made, was that instead of trusting the Great Law I trusted humans at the time when I had very severe glaucoma-like pain in my left eye. Instead of turning to the Great Harmonizer, I turned to a human surgeon. It is over two years now of suffering. I can only accept it and say, "My mistake, Thy will be done: Thou are just and I have to bear that which is just."

My mistake summarizes all the mistakes we have made and the great mistake we all once made. To find alibis would only complicate matters, because what is right is right, and what is wrong is wrong. The sooner we accept it and the sooner we realize that it is our own fault, the quicker we will be on our way to pay off.

Humans say, "Is there a God who permits those things?" Did we ask our Father, when we made the original mistake, for permission to make it? We went against the will of the Father.

We must think this matter over very seriously, especially with the approaching New Year, and even before then in these days of "Peace on Earth, Good Will to Men," when we commemorate the One who devoted all His life to bring the Message of Love to Humanity. He is now ignored. The Prince of War is now put in place of the Prince of Peace. Still there is hope, because of the Law of Evolution.

Some day we must awaken and take life earnestly. Humans try to break the last cord in the Lyre of Life, by refusing to accept the Law of Hope. It is not an easy problem, but the sooner we learn, the sooner we will adjust ourselves to the Harmony that is underlying and overshadowing everything.

The Law of Evolution is working much stronger for those in misery than for those in relatively good condition.

We should enjoy everything in the proper way and at the proper time. Do not forget those who are suffering. Do not forget that there are some who are friendless and lonely. It will make us more gentle in heart. It will make us realize that the world is in disharmony. By trying to sympathize with them, we help the Law of Evolution to work through us.

Hardness can only endure to a certain limit. Those who suffer, those who cry, will have their tears dried up, because they have paid.

Evolution, when understood, is our first awakening. It is a wonderful Law. It is a law that fulfills that promise of "Peace on Earth, Good Will to Men." Let us try to cultivate that Law in our minds. Then that Law will help us find the right way. The Laws of the Universe never rush. They grind slowly, but strongly.

No thought, word or action is ever lost. Some day it will come back to us, and we will have to pay. The coming year will not be an easy year. Therefore, we should prepare ourselves as well as we can. This does not mean that we should whine about it. What's wrong is wrong. If millions deny the trouble that will not remove it. It must mentally, physically and emotionally be removed. Try to realize the very opposite of the wrong which is besetting Humanity.

Our Organization must be militant. This means to be active in a constructive way. Deny on the mental plane that the trouble has any power over us. Do not make a statement that the trouble does not exist. This would be only the statement of a blind person. In the Realm of Eternal Harmony it does not exist, but in this dream through which we are passing, it has become a substantial reality. Alibis will never change it. Say, "Get behind me. I do not want anything to do with it, unless I can remove it."

We should never try to help others unless they ask to be helped. We should never be friendly to an enemy unless he asks to be friends. People may not be able to come to us physically, but let them come mentally and emotionally all the way to us.

We make so many mistakes because we do not use our Higher Self and its wisdom to guide us. We take life too easy. Happiness not earned is not enjoyed. We must work hard to enjoy happiness. This is a law. No special dispensation can remove the law.

Lesson Eighteen

PEACE AND HARMONY

"I am All Power. With what Power then Art Thou Fighting Me." This statement is the most fundamental rebuke to our human mind – the deserter, the one who took the place of the rightful King, The Divine Intelligence.

This lie of the human mind has to be exposed. The lie tried to take the rightful crown belonging to Supreme Intelligence. This has to be challenged, challenged all the time. This is a very tiresome, continuous fight. Oh, my friends, how little we know what fighting means! Whatever we do we have to put effort into it, and that means to fight.

"I am All Power," says The Eternal, the Supreme Intelligence governing the Universe. Yet humans fight that Power. They fight Life with the gift of Life. They fight Matter with the gift of Matter. How inconsistent humans are! Why? Because they forget that wonderful statement, "I can do nothing by myself, but with the Father I can do all."

Mind, the bearer of Eternal Light, how low has it fallen! It has fallen from the high estate of Harmony to the lowest state of Hell. It wants to create a new heaven and a new earth. With what power will that come into existence?

The power of All Power was perverted, but some day it will free Itself.

That Power, which we should have turned in the right direction, is causing all our trouble. It is the result of the Sacred Power within us breaking the fetters of ignorance and fear. When fetters are broken, it causes us to suffer. It is very unfortunate that we need to go through the process of burning dross in order to have the pure gold redeemed. But that is the way we made our own life. We built a big structure on a foundation of sand, both collectively and individually. We are now paying the price collectively and individually.

We should use All Power, which is always at our disposal, to fight for the right cause. We must make an effort. The All Power helps us to fight that fight. There is always that Ever Present Help. That Help is always here just like the help of air to our life and unfoldment, just like the help of food to unfold us, just like the companionship of people who are channels for the Eternal Help. If every human forsakes us, if every human help seems to be of no avail, there is always One Help, closer than human help. We must train our minds, in order to understand better the practical value of these lessons on Mind.

Wisdom means a broad view of the whole situation, and an understanding of the situation. Wisdom has been considered from time immemorial as one of the most valuable traits of our mind. We, individually and collectively, suffer now from a lack of wisdom. We should ask to be open to the wisdom of The Eternal. Wisdom penetrates deep into every thought and action. Solomon, when asked if he wanted wealth or a long happy life, or wisdom, chose wisdom.

Wisdom has a great value in our life. It comes not from our subconsciousness, but by trying to incorporate into our consciousness what our Higher Self endeavors to tell us.

The wiser we become, the more harmonious we become, and the more useful we are to ourselves and to others. Wisdom governs justice and includes Love. Wise people are fundamentally kind people. Wisdom never forgets anything, but never resents anything.

When one cannot help something, wisdom says, leave it alone. Some people don't want to be helped. They must be left alone. Some people want to be helped, but don't make the proper effort themselves. They must be left alone. Others really want to be helped. Do not throw pearls before swine.

People say they try to be good but don't succeed. They lack wisdom. To be wise means to be deep, but sometimes very unassuming on the surface. If we care for a person, or a group, we should exercise that much more discrimination.

The road to Wisdom, though a straight road, is not an easy one. There are steps to it. The first step is attention. We should notice the material, mental and emotional things we come in contact with. Develop the ability to notice things. We cannot notice things without developing attention. We cannot do this if we are half asleep mentally. The more we are awake mentally the more we are approaching wisdom. There is nothing worse than to walk on clouds. Our human body is much too heavy for such an adventure. People thinking of higher spiritual things think they can disregard the material. It must be 50-50.

The foundation of wisdom is the FourSquare, which is also the foundation of everything, even of the Eternal Father.

Wisdom, though of a spiritual nature, is the most important mental manifestation in our present life, because it not only governs the mental plane, but penetrates deeply into the physical plane.

Wisdom is above mental walls and regulation. It is a Law. Laws are manifestations of the Eternal, which cannot be broken or removed. Regulations are man-made to manifest certain aspects of the Law. There is no need of regulation in the Realm of Harmony.

In human life, where we do not perceive wisdom in its whole aspect, we have regulations. These are usually to prevent people from doing the wrong things rather than to inspire them to do the right things. Regulations are man-made and anything that is man-made can be changed.

Jesus did not manifest wisdom when He lost His temper and drove the money lenders out of the temple with a whip. He was not a perfect being. No leader is perfect. People can follow a leader in spite of his mistakes, but they should not be blind to his mistakes. Any fool can obey an order, but it requires intelligence and wisdom to disobey them. I try to not make mistakes, but freely admit that I often do make them.

Wisdom is not shouted from the tops of houses. It is in the Tower of Silence. We should tell people personally their mistakes, but publicly try to bring out their qualities. There is always something in each of us that can be brought out. This is a rule and not a law. Sometimes shortcomings, under certain conditions, have to be exposed. This is often so in politics.

Our Conscious Self is in a peculiar situation. On the one side we receive wisdom from our Higher Self. On the other side we receive foolishness from our subconsciousness. The most manifested characteristic of subconsciousness is foolishness. Subconsciousness is a mistake in itself. Wisdom tells us we made a mistake. Subconsciousness never acknowledges that it made a mistake. One can never reason with subconsciousness. The majority of people, being slaves of their subconsciousness, are very foolish.

We were all very foolish when we originally fought the Eternal and fell to this plane as a result. This was the most unbelievable act of foolishness. In our pride, we blew up in a fireworks of conceit. Not all of the beings followed Lucifer (mind). Those who profess to be exact in figures claim that one-third fell and two thirds remained loyal to the Eternal. This proportion is not known to be correct.

That fight has not the grandeur portrayed by Milton in his “Paradise Lost.” He made Lucifer a very tragic but grand figure. I feel that all Lucifer deserves is to have a fool’s cap, or dunce cap, put on his head and bells on his clothes, the bells ringing and saying, “What a fool I am!”

I think this is the first time the Devil has been made ridiculous. Religions made the Devil horrible and destructive. No one can ever respect anything that is ridiculous. There is nothing more destructive than to ridicule something.

What does this all amount to? We speak with a cold, analyzing mind. We have no right to analyze Love foolishly, but Mind should be analyzed by our reason. It is foolish to say that mind is untouchable.

In our Conscious Self we are now facing our Higher Self, or wisdom, on one side and our lower self, or foolishness, on the other side. Unfortunately, foolishness usually wins.

It is easier to look down than up because we are earthbound and fear, being one of the most important factors in our life, makes us feel that the ground is not safe. Since this has been carried by humans and animals for millions of years, we always look down. Therefore, it is quite human to look down. The sky has vibrations, and so does the ground. The Superconsciousness and the subconsciousness are the same way. Whichever way we turn, we get the vibrations from that direction.

To pray with heads down is entirely the wrong way to pray. The American Indians never prayed with their heads down. The first settlers in the United States should have learned this from the Indians and use it, but they did not.

Humility is not expressed in the position of the head, but with the eyes. We should look upward and say, "Father, help me."

When we turn our gaze to the Higher Self we get wisdom, when to the lower self we get foolishness.

Wisdom is of very few words. Foolishness is of many words. Chattering is a sign of foolishness. Monkeys continually chatter. People who jabber use monkey talk.

Today we see the reign of foolishness more pronounced than ever before because subconsciousness is trying to rule the world.

We should say the right thing at the right time. It is not necessary to give thousands of explanations.

When we hear someone retract a statement, we should try to find out the reason for the retraction.

Superconsciousness battles with wisdom and subconsciousness with the darts of foolishness. This accounts for the great turmoil in the world today.

Subconsciousness in the minds of humanity is a very powerful weapon. No individual who does not know how can rise above it. A few individuals are doing it, but a unit alone cannot affect about two billion people. An individual is able to tune in on the intelligence of the Eternal, or on a group which is feeling the same towards the Eternal.

A mental avalanche is a grouping of individual minds working in a certain direction. It is easier to work on a downward than an upward direction.

The majority of people have little discrimination between right and wrong. They take for granted a wrong thing that is varnished with a little truth. We should judge a thing broadly and deeply. There are only a few people who are deep minded.

An immediate and colossal problem is "What is right?" Truth is ideal right. Let us try to find right as much as possible, if not ideal right.

People above the masses have reached extraordinary results in certain lines. The more we find of what is right, the less there is of what we once thought was right. There has never been a time of so much upsetting of various scientific ideas. The majority of scientists do not know what science really is. Facts are so unstable that the more we unfold our knowledge, the less we understand the facts. Mathematics is an exception, but the riddle of Life cannot be expressed in mathematical formulae.

The minds of people are not working deep enough to solve the problems of War. They do not understand that the principle of war is conquering, not killing. The skunk protects itself and conquers its enemies with its odor. Nature thus proves that we can protect ourselves without killing. We who claim to be masters of the situation should be able to do as well as the skunk. The majority of people hate to kill other humans. The desire for killing has been with people for millions of years. They try to get rid of it by hunting and fishing.

We, as individuals want to fight for what is right. We should do this on an individual scale before attempting it on a mass scale. We do not know the difference between right and wrong. We should consider the group surrounding the individual. We ignore others, either deliberately or unconsciously. We must try to be wise. Humanity needs this above everything else. The power and need of wisdom is best expressed in the Baghavd Gita.

Without our knowing it, Superconsciousness plays a more important role in our lives than subconsciousness because it is always watching us and reminding us when we make a mistake. Subconsciousness knows no right and so cannot do this. Why does Superconsciousness give us such good advice? Because it is wise. Wisdom is the product of the operation through us of our Higher Self. When our subconsciousness works, it is like an explosion. It has a sudden urge. It does not know patience; yet sometimes we have to be impatient. Wisdom consists of using every possible means to fight and win the Battle of Life.

Usually impatience causes us lots of trouble because it does not permit us to reason things properly. To be patient with an unbearable or intolerable thing is foolishness. If there is a hope that a thing can be adjusted, patience is a very necessary way to handle and conquer the situation. It requires discrimination.

We have to be patient in that we cannot suddenly change the events of the whole world, but we should be impatient that world events do not take away from us the decision of what is right and what is wrong.

A very desirable advice from others can sometimes start us thinking along lines where we had been thinking weakly. The majority of people do not listen to advice, but are affected by the general mass thinking. We should learn to follow only the influence of our Higher Self. It helps us to gain mental poise and stability, so that no outside influence can affect us.

We cannot change our perspective of life in the twinkling of an eye. There are continual changes and adjustments in life. In these adjustments we should try never to lose our fundamental ways of thinking. A person who is fundamentally honest, when surrounded by crooks and liars, will think, feel and act honestly.

Humans fundamentally have fine traits, but these must be brought out, sometimes through many incarnations.

1. Some people are energetic. Let them manifest energy constructively.
2. Some people are developed mentally. Let them manifest mental qualities constructively.
3. Some people are honest and sincere. Let them manifest honesty constructively.
4. Some people are emotionally developed. Let them manifest Love constructively.

We can always manage our life so that we can manifest it along these four lines. If we disobey these, confusion is manifested.

We have complicated regulations because most people do not follow their own higher guidance. Today we are letting ourselves be guided by the world way of thinking, which is at a very low ebb. This has been brought on by the people themselves.

It is time now to awaken ourselves as much as we can. We must be awakened to the increased smoke coming from the battle between our Superconsciousness and our subconsciousness. The battle rages usually on the mental plane, but affects the physical plane.

We must more and more try to be as wise as we can in every direction. We cannot be foolish any longer. It demands too high a price. Wisdom never demands a price. It gives a reward, sometimes a very far-reaching reward. People say they are willing to pay the price. This shows how foolish they are.

We are facing hard times as a result of our foolishness. If we ask, is it worth the price we are paying, we have to say, "No" from the physical, mental and spiritual sides.

Usually foolish people are nice, pleasant people, but there is no substance back of them. It pays to be wise. We have such a peculiar, wrong attitude towards wisdom, our Higher Self, and so forth. We are not to be straight laced or puritanical. We may have different concepts as to what constitutes the joy of living. Those people who are dissolute are foolish. Those who say they are going to limit themselves in every direction are foolish. Foolishness is a very subtle trait of our subconsciousness.

The best way to rise above foolishness and to build a raft of wisdom is to be sincere. Be oneself. We have pretended too much. We have lost our sense of our own sincerity and our faith in the sincerity of others. We are living in a world of doubt and hesitation.

The Higher Note in our nature is very strong, especially in the Lightbearers. We should take advantage of that innate condition of our mind – Sincerity. This has on one side, in the FourSquare, Mind and on the other side Emotion. It is based on emotion, which is a ray of Love. We must be true to our Higher Self and to the fundamental conditions of the FourSquare. Be genuine and true to others. If we do this we will finally be true to Supreme Power, which will be our protection. As we are mortals, we sometimes slip, but we can pick ourselves up. If the whole world would act in this way, humanity would improve by leaps and bounds.

We cannot be sincere on the bigger scale and in the deepest way unless we learn to be sincere on a smaller scale. Evolution cannot be reversed. We start from the smallest and grow into the larger. This is clearest expressed in mathematics, when one progresses to 9. This is the problem facing humanity. It always has been so.

When we are small children we are usually sincere. But, due to outside influences, we lose that very beautiful trait of sincerity. When we reach a certain age, instead of unfolding, we go downward. When we become old, we begin to realize slowly the value of sincerity. Older people, no matter how sophisticated they are, usually become sincere. We should not wait until we are old to come into our own. If we do so in Youth, we are able to see and enjoy life to the greatest extent. But we are not educated this way. This affects our everyday life. We fail to get the most and best out of life.

Wisdom is supposed to be not only the attribute of old age, but also of youth. Youth usually lacks wisdom. Life is not meant to be lived with the idea of losing, but of winning. One should win on the three planes – physically, mentally and spiritually. The value of wisdom when we are young is that much more important than when we are old, in that we can more quickly advance and make the world advance.

We cannot individually much affect the whole for this world, but we should do the most to affect ourselves in our unfoldment. Life is misdirected for almost all of us. We should do something to prevent ourselves from being carried away by world conditions. We should look up to something better and deny having anything to do with or participate mentally with that which is wrong. This requires that we go against the current of humanity. We can do this if we have the help of our surroundings.

Humanity is like a big pool and each of us is a drop in it. Which is stronger, the pool or the drop? We have never known that the pool can take away the color of the drop, but the drop can color the entire pool. One drop can upset the entire pool. One individual can resist the whole of humanity and affect the whole of humanity.

The elements on this planet are to a great extent affected by the thinking of humanity. We are much more important than we think we are. Jesus was able to still the waters of the Sea of Galilee. People think they can, with their machines, master the earth, but they cannot master a wind as did Jesus.

We must learn to understand things beyond our senses. We should try to avoid every possible delusion or illusion. The whole of life today is an ever increasing illusion. One of the worst ways to create an illusion is to be insincere. Insincerity is a lie. A deliberate lie will not delude anyone. When we have reached the fullness of wisdom we suddenly realize that something is wrong. We should do something about it.

Consciousness, guided by our Higher Self, is willing to admit a mistake, but subconsciousness is never willing to do so. It tries to find every possible alibi. An alibi is a way to avoid telling the truth. Its main aim is to give more life to a lie. A truth does not need an alibi.

Insincerity is so terrible, because, being so in little things, we will also be so in big things.

We can feel insincerity in a handshake. There must be Love in a handshake. Love is the strongest power there is. We should not assert our sincerity in a weak way. Weakness is unforgivable in human life.

Humanity is less and less sincere. The more this is so, the more we are going to pay the price.

The devil, in religions, is sincere about torturing the souls. The good souls, looking down, applaud. What an unbelievable insincerity!

No human can ever camouflage the expression of his eyes. We cannot lie through the eyes. Most people judge by the general expression of the face, but this can be controlled by the mind. People cannot control what shines through the pupil of the eye. The pupil is the lens from the unknown of the inside to the known of the outside. It is the only place where sincerity can be found manifested in a physical way.

We want to be good friends, but we don't know what it is to do so. We should try to be ourselves, but from a constructive basis. Smile at what is worth smiling at, but not when it is out of place. A smile is out of place when things are sad or tragic. It hurts. We must learn to be sincere, without hurting people. We should be sincere in trying to make them feel good. Extend a hand to prevent people from stumbling, even mental stumbling. We must have natural sincerity. One needs support at times when there seems to be no support.

Sincerity is the connecting link between what we think and what we say or do. Our thoughts are always sincere. We cannot pretend in our thoughts, but we can in our actions or sayings.

An act or word which does not correspond to the thought back of it is stillborn. A thought which has not been substantiated by an act or word has not been incorporated into life. The act is the final incorporation into life. The act must be according to the thought. Some people act in a very nice way which does not correspond to the thoughts back of it. They are hypocrites.

The thought is the cause and the act is the effect. The present world conditions are due to the ever increasing contradiction between thoughts and acts. Explosions of a destructive nature in thoughts later materialize on the physical plane.

People talk more than ever about constructive action of our mind, but truth is less prevalent than ever before. Our conscious self is more open than ever before to the influence of subconsciousness. We should be more open to Superconsciousness, but people are not looking up.

Humanity has been dreaming away life for countless generations. For most people the dreams did not come true.

When conditions became almost unbearable, the religious people shifted the expectation of some happiness into the Beyond. This worked to a certain extent on people who blindly believed it. Unfortunately, and fortunately for dreamers such a condition could not endure forever. Rational mind demanded proof. People are now less and less willing to accept things on faith.

The future does not build itself. We are now building it. This demands a sense of reality, of common sense. It demands a knowledge that if we start something wrong, it can never be made right. We must have faith not in human teachings or inspirational thoughts, but in an ability to perceive properly what life is presenting to us now, and how it will work in the future. The strange part with wrong things is that they start easily, but become more complicated the further they go.

When we submit to something wrong, our Higher Self always fights that which is wrong. If we do something right our Higher Self helps. If we are not inspired by our Higher Self, but lean instead on our subconsciousness, our Higher Self is still fighting. This wears out our conscious self. Our own consciousness wears us out.

Humans are much better than they seem to be and much worse than they think they are. They are better than they usually show themselves to be. In a moment of need, there is something in them that asserts itself. There is a continual fight between the two elements in our being, the Higher and lower.

Today things are getting worse. What can we do? To give up hope would be very foolish. But we should keep hope in the proper place in our mind. A stone rolling down a slope will only stop when it reaches the bottom. It is the same way with humans.

Little things which do not amount to anything accumulate, then press each other, then solidify. When accumulated sufficiently, a little concussion starts the whole moving down. When too many sorrows accumulate, an avalanche starts. It is the same way with constructive things. These accumulate and we also get a strong precipitation – a much needed rain, which will turn every barren thing into a green field.

Today humanity is reaping what it has been sowing for countless generations. It is a human calamity, a disaster. We should not try to stop the avalanche, but we should try to find a rock to which we can hang on, and not let the avalanche carry us to the bottom.

When the avalanche of life is dragging all of us, if we do not lose our presence of mind, if we do not lose hope, we usually get out, because our Higher Self, knowing that a Higher Power can help us, opens the way for us. It is not easy, because all we have is our inner feeling of that Power which cannot abandon us. When we need something very badly, and have faith that it will come, it does come. This has been proven thousands of times.

There is one great thing facing humanity today, and also each one of us. That is greed. There never was a time when greed was so manifested on this planet, working both in an open way and underground. Life on earth is not based on money alone. Money is a human interpretation of things.

The ruling incentive of humanity today is greed. Greed ultimately culminates in poverty. That is where the whole of humanity is heading now. Bankruptcy is facing all of us today. It has become a world-wide calamity brought about by greed. The United States will be the worst hit of all.

Back of greed is the fear that we will not have enough. Fear is one of the most destructive characteristics of subconsciousness. A greedy person is fundamentally afraid that he will be poor. A miser is rich only in fear. He keeps money without making the proper use of it.

Let us be generous materially, mentally and spiritually. Never disregard an opportunity to express it. Give to a stranger the richness of our smile. There are many ways to enrich others and not lose ourselves. There is always a way to help others. By helping others we help ourselves.

In our minds, always side with that which is right. Never mentally tolerate wrong. The moment we accept it, we become just as guilty as the originator. We should not blindly follow Government leaders. Our duty is to say we will not be a part of the greed of today.

How little we read in the newspapers today of constructive things! Why be such a pessimist and see always the destructive things? Why not look at the precious pearls amidst the destructive things? This can be done and should be done. We should also do this in our relations with other people. We should sympathize with them. Do not permit ourselves to be pushed down by the general condition of the world. Stick to the right and trust the Higher Power, and we will come out on top.

The conscious self is most important in our evolution, because it is the part which brings our final liberation from that which is not worthwhile. It started from a small dot, almost microscopical, and continually increased to its present condition.

Perhaps now is the most important time in our evolution, because we have reached a condition where we may be reaching maturity. Maturity means a balanced condition.

Maturity does not depend on age. Sometimes very young people have very mature minds, while other people, old in years, have minds almost like children. When the mind is mature it begins to see things which before it could not reasonable accept.

Reasoning must be three-dimensional. Reasoning between two points is superficial. There must also be depth. The value of mature reason depends on its depth. Depth of reasoning requires wisdom. Wisdom is usually a developed thing.

When we are in a condition of mature development of mind, we have continual problems facing us which cannot be solved in the twinkling of an eye. With developed reasoning we can reach conclusions in an extraordinary short time, almost like a flash. Most problems do not demand instantaneous solution. We have the problem of how to use reasoning alternately, quickly or slowly.

The Laws of Nature are set, but rules are not. Laws cannot have flexibility, but rules must be adjusted to the conditions where the rules are to be exercised. Rules may be rigid or highly movable.

A wise, mature mind has to figure out the pros and cons of any situation. Even a little situation demands a proper approach. The whole of an intelligent life is based on what we learn.

Our human mind is as cold as ice. That is why it is so difficult for us to warm up to anything.

Our discrimination must tell us when to do one thing and when another. A colorless life is almost not worth living. A skillful combination of colors gives an extraordinarily beautiful painting. Beautiful music is sounds skillfully arranged.

The present time has more problems than we probably ever had. The greatest problem, in facing them, is to understand how to stand in facing them. They must be faced in every direction. This always was and always will be the greatest handicap, until humanity has freed itself.

We are influenced by others. This is natural because we live a collective life. But within this collectivity, we should maintain our individuality. The development of an individual depends on what he thinks, feels and acts within himself.

When we are in the cusp of cycles there is always a hurrying to settle old accounts and enter into the new. There are smaller cycles in each generation. Young people are always in a hurry; old are slow. There is the fight between the old and the new, often an unconscious fight. To solve the problems requires the greatest wisdom anyone can have.

There is another peculiar condition. The world wants to be free. The sense of freedom is more expressed in the young generation. It is very difficult to achieve a very advanced freedom. The best step is to develop an inner freedom. This will later manifest without. True freedom is inside freedom.

So many people think they can be free themselves by being afraid to face life. This is a complete misconception. Fear has a paralyzing effect on humans, and even on animals and plants. The idea that fear stimulates us is absolutely fallacious. Enthusiasm relaxes, but fear paralyzes. The greatest mistake we can make is to be afraid to make a mistake.

In present day conditions there is an unbelievable fear sweeping humanity. If people would realize the real condition of things today they would not want to live. Fortunately most people do not know. Hope is the last string in the lyre. People with even an average good health have no desire to do anything because of that terrible pressure of fear.

Some people with an understanding of life have a sense of security. Most people think only of material security.

The greatest security is in Love, not in Mind. Love may be colored differently but can never be anything but Love. Happiness is the result of having found a little love. When people have found a little love, they feel a little security. It helps them and carries them through difficulties.

Love and Mind are the greatest powers in the Universe. Humans seek security in Mind. Love must be wise and Mind must be loving.

In our present condition we are frozen. We do not know which way to turn. We must use very wisely our judgment, and be not influenced by the masses.

The voice of the people is not the Voice of God. It is the voice of the subconsciousness. The voice of the Higher Self is all that can be called the Voice of God. We should turn to it as much as possible.

More than ever before, we should try to free ourselves from outside influences. Each one of us is different. No other can identify himself with us, nor we with him. We can only partially see the approach of others to a problem. That is why when people are following their Higher Selves they come out better than when listening to others.

We should, however, ask advice, and give advice. People getting advice from more than one get conflicting advice, and the person therefore is more bewildered than before.

An individual asking advice has usually already made up his mind. If his decision is wrong, he will never accept right advice. If his decision is right, then right advice blends with it. No advice is accepted which is different from their own decision.

We learn our lessons in life either from experiences or thorough intuition. Straight reasoning is honest and simple. An alibi confuses. Lawyers complicate their sayings and try to win court cases through confusion.

People who are easily influenced do not know what they want. They have confused minds and are called rattlebrained. A constructive mind must be harmonious. It produces music and harmony. There is a connection between every note, and a meaning in each note. Rattlebrains have no connection between their notes, nothing but noise. We are not to be slaves of noise, but masters of music.

A rattle is all that a child wants. When it grows up, it wants better musical instruments, and it talks and acts harmoniously. In the present world we see nothing but rattling. This indicates humanity is still in its infancy. It is far, far from mature in its own mind.

Before leading others, one should be able to rule himself. He cannot learn to command before he learns to obey. When we have learned to control ourselves, we have learned one of the most difficult things. We must all learn to control ourselves. Then we will begin to be winners.

Humanity has established certain rules in order for it to progress with the least friction. We should not become victims of our surroundings, but should make the best use of them. Our present surroundings have frozen our way of thinking.

The rule of the masses is the most unbelievably impractical thing one could imagine. A stampede is the rule of the masses. They blindly follow an impulse. The individual is completely submerged by mass thinking. In a stampede, people do not discriminate whether a thing is good or evil. This is democracy in the truest explanation of the word.

When the United States was founded, George Washington did not belong to any political party. He was an aristocrat, and probably the first millionaire this county ever had. He was so wise he would not become a king. He was a true world citizen of the United States. His principles were worldwide principles. His advice to not get entangled in world troubles was good advice.

The ancient Jews had a very fine concept of Theocracy, or rule of God.

A democratic mind is a very confused mind. Confusion is in subconsciousness. Democracy is the rule of the subconscious. Autocracy is the rule of the Conscious Self. Theocracy is the rule of the Higher Self.

Democracy has made a terrible mess of the world. The democratic attitude will always produce a mess. We should treat everyone in a friendly way, but everyone in a different way. We should use discrimination in everything we do, and towards everybody.

An autocratic condition of the mind is a step up. Such a person relies entirely on his own interpretation of things.

Theocracy is government by Intuition, by the Eternal. Through Intuition God speaks to us. Not one of us can do one thing right without the help of the Eternal. If we do a thing right, we do it with the Eternal. If we make a mistake, it is due to some undesirable influence. We are not under the guidance of our Higher Self. If others make mistakes and we do not notice them, we are just as bad as they. We do not have clarity of perception.

We must decide under what government we want to be. We cannot suddenly jump out of democracy into Theocracy. We must go through autocracy, where we rule ourselves. In autocracy, we stick to our own principles, which we have found to be right. In Theocracy, we say, “Not my will, but the will of the Eternal shall be done.”

We have never learned anything in a few minutes. We must pay attention, which is the easiest way to learn. Mentally we are now children. There is no age limit in learning.

Autocracy always looks up. It is based on the principle of Growth. Democracy is based on the principle of spreading, as a drop of oil on the floor. Theocracy is four-dimensional.

Today we must claim for our self an autocratic attitude, but influenced by our Higher Self. We can reach this stage if we take the proper steps in the right direction. We should stand for our own right of thinking. This does not mean that we should refuse to blend with others. In the present condition, where there is so much wrong thinking, we should not be influenced by it. We should stick to our own way of thinking. If the thoughts of others blend with ours, we should welcome them, as they will strengthen our thoughts and our thoughts will strengthen those of the other people. But we must use discrimination. On this is based the whole progress of Evolution.

We are in the midst of Eternity. It will always be this way, but this is very difficult to realize.

(At this time in the lesson, Svetozar had the entire class spend about five minutes thinking backwards through eons and eons and spirals of time until we could go back no farther. Then we spent another five minutes thinking forward through many spirals of Evolution until our minds became blank, and until we could hardly endure the thought any longer.)

The reason for this experiment was to show us that we have still more unfoldment in the future. In the past, we had unfoldment in spite of all the mistakes we made. Sometimes it takes an entire lifetime to realize a mistake. Realization is proof of our unfoldment.

We should realize that life must be fundamentally harmonious. If not supremely harmonious we could not bear it. If fear comes, our subconsciousness is afraid of its own self, because it realizes that it is wrong. Since it is limited, it throws away all indications of what it does not want. It is a cradle of death and of birth, and wants to be limited. It is limited and tortured by its own fear. It cannot bear the idea of Eternity.

Mental pressure is usually due to some mistake we have made, but is sometimes caused by physical sickness. Each mistake is simply walking on the wrong path. Our Higher Self realizes the mistake. As soon as we acknowledge the mistake, we feel a relief. The moment the cause is destroyed, we can adjust ourself.

If we did not have shortcomings we would not feel the world pressure. We would only see the world conditions. An irritable person can be more affected by outside conditions than a peaceful one. We must see things clearly but not be affected by them.

Group thinking helps evolution, but we should not accept things blindly without reasoning them out. There are so few things right that we should fight all the time. If we want to be on the line of unfoldment, we must more and more stick to what we think is right and not give in.

If we think someone else's thinking is wrong, we should ask ourselves, "Why do we think it is wrong, and why is our way right?" A standard of right is that which does good to a person in an ever continuing way. That which profits us at the expense of other people is wrong. If someone opposes strongly something that is wrong, he is helping humanity's evolution.

Every intelligent and well-wishing individual wants to be on the side of those who disagree with masses. This is for the good of the masses.

It is very difficult to keep one's mind honest when working in certain branches of activity, such as economics, for example. A person using advertising, as it is now, cannot be honest. We are compelled to work with lies, but we should know that they are lies. Then we are not affected by the lies.

In a world of so much dishonesty, it is very difficult to remain honest, but it can be done. If we do not see this, we will be absolutely submerged in the wrong way of thinking. Soldiers who kill, do so, not because they wish to do so. They are only channels for the thinking and the orders of others. It is very difficult to disapprove of a thing, and still be with it. The moment we disapprove a thing because it is not according to our standard, that moment we are a winner.

In the world of untruths, we should stick as much as we can to what we consider to be truth, to be right, to be good. We must see that whatever we do, we should never be a loser, because sacrifice is only right, when something is wrong. When we sacrifice what is wrong, we gain that much. If we sacrifice something which is for our own good, we lose out. It is a misconception of what is right and what is wrong. What is right benefits everyone. If it hurts someone it is wrong. It is hard for us to understand what is right and what is wrong. The moment we begin to lose, there is something wrong.

More and more now we are forced to face life, not as we dream about it, but as life presents itself to us. We have to pay the price for a mistake. There is no use to worry about the mistake, but we should try to correct it. We are confronting life every moment, and we are likely to make mistakes every moment. Lightbearers have the advantage in that, with what we know, we can counteract mistakes by outbalancing them with something right.

The whole mind of humanity is in a most irritable condition, and is getting more and more that way. We are irritated when we encounter opposition. We are living in a world of disharmony, and those who want to be harmonious encounter the most opposition. This is true for the time being only, but no one knows how long it will last.

When we do not win on the surface, we unquestionably win a great victory within. We grow closer to that which is right.

Opposition usually starts slowly. In time it becomes so strong it stops our progress. This is the human road. With the Divine Road we usually meet great opposition to start with. This then begins to dwindle and dwindle until the road becomes easy. The human broad road becomes so narrow it crushes us. The Divine Road starts so narrow that we wonder if we can go through it, but it grows broader and broader. Some beings start in a very quiet, unassuming way, but as they move on, they broaden out and grow.

In this world everyone who tries to think rightly will meet a tremendous opposition. This is called world pressure. The wrong of the present day conditions affects much more those who want to do the right thing than those who want to do the wrong thing.

There should be no sacrifice. It should be viewed as the proper step to growth. When we plant a seed, we must dig the ground. This is seemingly a sacrifice of time and energy, and also of the seed. If this is done properly, and if we later take the proper care of the growing seed, we are bound to have a good harvest. This is not a sacrifice, but an investment.

We are to invest ourselves in life. We should look up at life in this matter-of-fact way. If we put much into life, we can expect a good investment. This applies to the material, mental and spiritual planes.

Jesus coming to earth invested Himself in a human life and for a certain purpose, which was to make life better. His investment was not in vain. It was a good investment – materially, mentally and spiritually.

People who have done some good have invested themselves into life. If we invest ourselves properly, we can in turn make proper investments. Many things which as ideas are excellent are wrong at the wrong time. We must be careful how to invest ourselves. Investment means to put ourselves into something. Each act is an extension of ourselves. We cannot be disassociated from life and expect to succeed.

Our innate sense of balance tells us what is right and what is wrong. We should not try to find an alibi for something that is wrong. A wrong thing cannot be made right by an alibi. That which is right does not need an alibi. There is never a way to whitewash something that is wrong. We must be vigorous and not do foolish things.

Will we succeed? Why not! There is no reason why we should not. Where is the secret of success? Within each of us. It is to realize more and more our Higher Self. Sometimes we learn this by hard experience. Sometimes it comes in a natural way. But we must learn it. We must look out of this material world and into something higher. Let us be stimulated and guided by a Star, no matter how dark the night. If we have ever seen the Star and remember it, we can never lose our direction.

What is right we should engrave as deeply as we can in our minds. There are lots of fine thoughts and fine actions by fine people in this world.

We should not be praised if we do all that life demands. We should be satisfied that we had the opportunity to do the right thing and should be grateful for having done it. All of life is a series of opportunities.

In order to more easily overcome obstacles do not say they are a hard thing to do. Say, "With the help of the Great Law, the Great Law will make it easy." The Great Law always works for us, but we do not adjust ourselves to it. We do not work in tune with it. When we lose in any direction, it shows we did not coordinate ourselves with the Great Law. If we are beginning to win, even a battle, it shows we are cooperating with Life.

The whole of life must be lived as a conscious life. It is quite true that the whole of life is a dream. But once we started dreaming it, it became to us a reality. We must awaken from it. Each time we do the right thing, think the right thing, discern the right thing, we are that much awakened. The final awakening will not come for millions of years yet. Nothing happens suddenly.

Things are moving so slowly and gradually we can hardly notice it. We should acknowledge that things are getting worse. This is proof that we are ridding from our subconsciousness many of the things that are in it. Subconsciousness is not the guiding instinct to do the right thing, as psychologists teach, but to do the wrong thing.

Many things we must accept, but not endorse. People do not understand the difference in the meaning of words.

The Word of Nature is Manifested Law. All Laws of Nature are Words of Nature. Everything created on the earth is created according to Law. Everything in this world must be done according to Law.

The bad things in subconsciousness come out through the process of Evolution. In subconsciousness they are like latent diseases until something brings them out. We have in our body the latent germs of all diseases. Stimulation brings out some particular germ, and we have that disease. It is the same way with mental germs, but they are worse than the physical germs. In a world where there are so many mental germs thrown out, it is a miracle so many people are still alive.

There is no way to escape from any physical, mental or emotional problem. We may ask, "Why live under such conditions?" The answer is, "We cannot help it." We should look up at life from an ageless viewpoint.

Do not permit people to put us into a certain class. We are each an individual and are in a class by ourself.

Lazy people think they have not enough time to do a thing. Active people always have time. They achieve much.

Because of mental confusion, we forget things when we are in a hurry. We should use judgment and decide whether we must do things slowly or with quick movement. We should direct our pace according to our needs. We must do things rightly. We should never be in a hurry, but we can be active. We should do everything we do in the FourSquare way.

Mind is not the governing power. Love, or the Power of Attraction, is the governing Power. We are in a mess today because we did not use mind properly. We are getting more and more unbalanced. This will bring about a condition where we will look around to find something to which we can hold. Humanity thought that religions would do, but - no. Some thought that law would do – no. Some thought that persuasion would do – no. Everything that has been tried so far has failed, because everything was lopsided.

In times to come, for millions of years, nothing will be more important than to tell people to be balanced. They must not be spasmodic, but continually balanced. We must be balanced physically, mentally and emotionally. No matter how terrible conditions will become, we must not become more unbalanced, or crushed by it.

Never admit that any problem that comes to us cannot be solved.

If we try to do everything rightly, we will in time have to pay attention only to the big things. The little ones will come automatically.

This is the way to get as much peace and harmony as possible. The more we are balanced, the better it works.

We, as Lightbearers, are not to try to develop ourselves into supermen or superwomen, but just good human beings. If we do the best we can, we will be protected in spite of the raging storm. No one knows how long the storm will last.

When we win over our own shortcomings, it is the greatest satisfaction. Within us we find the peace and harmony which we try to find some other place. Why look for happiness far away? Learn to perceive it, where it is, within us. Every real Friend of Humanity and every Great Teacher comes to this same conclusion.

Lesson Nineteen

IN QUEST OF THE UNKNOWN

The whole of the Christian World is today commemorating the Crucifixion, the final act in one of the greatest tragedies for Humanity that ever took place on this Planet. Is it any wonder that we feel tonight a greater pressure than usual because there are many millions who still follow the Teachings of Jesus and give due respect to the Teacher.

For all that Jesus tried to do, the reward was a crown of thorns on His head, four nails in His hands and feet, a thrust of a spear in His side, a drink of bitter gall on a sponge, and a mockery as a song in His last dying moments. That was the reward He received from Humanity.

There was that betrayal, that lack of gratitude to one who was all gratitude to his Father for being permitted to come to this World to help Humanity. Mentally and morally He died on the Cross but probably not physically.

That was a great tragedy which we not only commemorate, but for which we pay a great price today. The Jews were once one of the great Nations on Earth. They had a well developed civilization, one of the most outstanding rulers, King Solomon, and they produced more prophets than any other Nation. Now, for about twenty centuries, they have been the most downtrodden and despised race. Some individuals among them have been much better than the so-called Gentiles, but collectively there is a curse on them.

Why are they despised? Pilot gave them three chances to vote for freedom for Jesus. He then said, "I wash my hands." The people whom Jesus healed, fed, who shouted "Hosanna, Son of David," and who spread palms on his path, became his worst enemies because there was no gratitude in their hearts. They said, "Let His blood be on us and our children." This curse has now lasted for twenty centuries. They put into operation a Law, and they must live through that Law. Probably every one of them are living now on this Planet. Perhaps some of us present here tonight are those people. We do not know. It shows what a terrible price we pay for words which we sometimes do not understand when we utter them, and especially for ingratitude. We should be grateful, especially if we get a little good in a world of so much evil.

Present day conditions show that there is an unbelievable amount of ingratitude which we have not acknowledged. If all of this were exposed, probably one-half of the people in the United States would be in prison for misdeeds. Yet we think we are doing the right things.

I realize that THE LIGHTBEARERS have forgotten the Guiding Power of Love. We wanted everything strong mentally and scientific. What do we know of science? That is why it is so difficult to find Right. We must be guided by the only Power that can show us Right, by Love, a deep, wide, unassuming feeling which is the foundation of Harmony throughout the whole Universe. Even though we are not religious, we should understand the one whom religions proclaim a God. All he wanted was to be a decent Friend, a friend in trouble. He never imposed himself on others. In the story of the Lepers only one came back. He showed that he meant more than the verbal discharge of a debt. Words must be backed up by acts.

Everyone who tries to do something for Humanity, is included in the Crucifixion of everything Right on the Cross of Life by those who do not know what is right. The Crucifixion will remain a blot on the whole of the race. It spreads like a contagious disease over the whole of Humanity. Everyone who does not do the right thing, crucifies his Higher Self. We are paying the price all the time for the crucifixion of our Higher Self which we disregard, betray and crucify.

That last page in the life of Jesus was not the last page in his life. His Higher Self was not gone yet. It remained in a kind of suspended condition on earth.

The Easter Resurrection is the Hope of Humanity. No matter how low we have fallen, it is not hopeless. The rose may be cut from the bush, but it is still alive and fragrant. A fire on the altar may be almost extinguished, but it still has life. When the strings on a lyre break, the notes are still vibrating and carrying a beautiful message to Humanity. We should not say we are dead. We are still alive.

Sunday we will be reminded of the reappearance on Earth of Jesus, the Symbol of our Higher Self. We must not give up the hope of the reappearance in our life of our Higher Self. That which is undying can never die. That which is called Good or God, in us will always function here and after we have discarded our earthly shell.

Through the experience of Crucifixion in a broad way as Science of Being explains it, we should find not only consolation, but stimulation to bring out that which we call dead, and use It again. That stone which is closing the Higher Self, that stone of our unbelieving ignorance, stubbornness and unbalance, will be rolled away and we will see our Higher Self resurrected. We must be burned down in that limited condition in which we now live, to be reborn again strong and useful.

That Unknown in us. That which is the source of everything. That which is unborn and never dies. That from which we came and to which we return, is our Higher Self. It has no end of Powers and Qualities, each to be brought out. That is the Great Unknown in us, the latent forces to be dug for, to be brought out, to be recognized and used according to what each is to be used for, the wonderful Thing which is, with the help of the Great Law, with the help of our mind, to tune in and bring out for our own good and for the good of those with whom we come into contact.

The problem of the Quest for the Unknown started because of ingratitude. If we had been grateful in the Realm of Harmony to our Divine Mother-Father, if we had listened to Love, we would have all the Good which makes our life as beautiful as life should be. We did not listen. The pride of our mind said, "No, I do not need help, I am going to find within me all I need. Gratitude means weakness, I am strong."

What a price we pay for ingratitude. A most terrible price. For one moment of ingratitude, we have been suffering billions of years and will continue to suffer. In the Realm of Harmony, one moment is like a thousand years. The length does not count, but the magnitude of the mistake is very important. Since we made a mistake to the Eternal, the punishment should be Eternal, but it will not be. Since it started, it must some day end. Since we faced the Eternal with ingratitude, the penalty will be long, long, long.

The greatest quest was started the very moment when Humanity was born on the Planet as human beings. Why? Where? When? What? These questions are infinitely greater than what we should eat, wear or how we should live. The answers would give us enough wisdom to solve all other problems. When Humans will sing and shout Christ is Risen, let us try to remember that some day the Higher Voice will sing in our hearts, "Yes, out of the ashes am I reborn a New Being, a better Being, a finer Being, a beginning of something more worth living."

The most important question Humanity has ever asked of itself is, "What is Life about? Why are we here?" Primitive people could not intelligently ask the question, but it must have been deeply buried in them. It must have expressed itself in the words, "I wonder." Children not yet old enough to talk, have the question written in their eyes.

When we are grown up and puzzled by the problems of life, we sometimes sit and stare, not thinking of much of anything. Yet if one looks into the eyes of such a person at that time, he can distinctly see the same question. When one reaches the end of his earthly existence, when he is about on the point of crossing into the Unknown, there is again the same expression in his eyes. There must be an unbelievably important reason why we ask this question unconsciously at the beginning of life and consciously at the end of life.

We know that although we are children of Nature, we are discontented with Nature. One reason for this is because of the unbalanced condition of the mind. The mind should be developed in the same proportion as is our Higher Self. But because of extraordinary achievements, especially mechanical, people have been led to believe that Nature is subservient to us. In other words, that the child dominates the parent. This is very noticeable in modern children. It must have been within us before, but is now coming into expression. Whether we express undesirable traits or desirable traits, they must have been within us before. Nothing can express itself from nothing. The seed must have already been there.

An individual's character can never be judged when conditions are harmonious. The only truly accurate understanding of people can be obtained when disharmonious things come to the surface. The surroundings effect an individual to the extent of bringing out what is latent within him. This is why we have all kinds of conditions affecting us, to bring out all latent things. The negative things brought out, help us to realize what is within. Today there is an unbelievable manifestation of undesirable traits.

It was prophesied that the day would come when that which is hidden would be revealed. We see this now. It shows up the United States in an unbelievable bad light. All we can say is that it was in the human mind millions and millions and millions of years ago. The improvement through countless incarnations was so slow because the question "Why are we here?" was never answered in a satisfactory way.

We are just a little piece of flesh with intelligence and emotion in it. Of all creatures the human child is the most helpless and unattractive, though the mothers see them through rose-colored spectacles of affection. The father takes a real interest in the child only when it begins to show some intelligence. To a monkey mother, it's her own child and she loves it. To love something which is very beautiful has no merit.

Why has the wondering about life been going on for countless ages? There must be a reason. We often wonder if animals wonder what life is about. No one can answer for sure. I am convinced that animals do wonder. I have tried to find the solution in the eyes of animals. Sometimes I have seen almost more than in the eyes of humans, especially if animals are sufferings. At their last moment of life, the same question, "I wonder" seems to arise. We must ask the question only of someone who knows more than we do. We cannot answer it ourself. We must look outside to get some answer.

The Why is actually a longing, a mental and emotional longing. Through unfolding of mental qualities, we try to satisfy the longing with the result that the more we think we know the more we realize that we know nothing. We are reminded of the Greek philosopher who said, "I know that I do not know anything." A little knowledge brings about a manifestation of conceit. Therefore the desire to solve the question through mental approaches has proven a complete failure. We realize we must ask, "Is there anything above us?" "Is There A God?" This is the only Power that can answer the question.

There is a tendency in us, back of the question in that seeking of the Unknown, to find there safety and protection. This started when Humans first appeared on Earth. It was the instinct within them that forced them to seek it, to seek the Higher Self through which the Laws of Evolution worked.

Humanity realizes that if we are to reach a high condition, we must have a high ideal within us. We seemingly lost this ideal, but we never did. We cannot lose that which is the fundamental part of us. The ideal in the seed conquers all obstacles and grows. Nothing could endure, unfold and reach its ultimate purpose if the ideal is not always in it. We do not lose ideals. We lose illusions. If it were not for ideals, life would be absolutely useless and hopeless for everything. The ideal is supposed to lift us up, to show us the goal.

We are here on Earth to multiply, not only physically, but also mentally and emotionally. The more we multiply emotionally, the nearer we come to the answer to the question, Why? When the question is answered we no longer wonder. We have the solution.

When the mental separation between us and the Eternal started, when we were hitting our heads against the Law of Harmony, that Law broke us. We are not victims of disharmony, because disharmony would have taken good care of us. Devils are always pictured as taking good care of their fellow devils. Like protects like. What brought misery into the World? Not evil, but the Law of Harmony. When Harmony is manifested in a world of disharmony. It is the worst enemy of those who are disharmonious. Good never helps evil. The Mission of Good is to fight evil to the bitter end. If we are determined to dwell in disharmony, there is no hope for us. But if we want to get rid of disharmony, then Harmony will help us. Unfortunately, humans enjoy little disharmonies, and do not do anything to rid themselves of them.

When we hit our heads against the impregnable Wall of Heaven, there is the inhospitable Gate of Harmony, there is St. Peter to prevent us going through. We cannot storm the Gates of Harmony. There is only one Key, a small key, the Golden Key of Love. A pure mind is much more important than a pure body. When we hit our mental heads against that which is unbreakable, we were broken to pieces. We are still seeking to bring those pieces together. We are ill because physically the pieces are wrongly placed together. There is not the proper relation of pieces to each other due to lack of cooperation. The head is the leader. Each part of the body refused to cooperate with the head. No part has been benefitted by this stroke. When we strike against Harmony and Cooperation, we are hit back and become disharmonious. If we break relations with Harmony, the price is terrible. We should fight, not to destroy disharmony, but to be on the same level with Harmony.

Humanity today is a very sick body because every part of the body is on strike now. The leader is the head, the brain. Love in Humanity is the heart. We lost the ideal of cooperation. Humanity since time immemorial not finding cooperation within, has tried to find it outside. The greatest cooperation is with the Eternal. We cannot have this unless we start with the lower. We have not yet learned to be decent with that which is on our own level. We try to reach something far beyond. We should start with simple love. The more we add to it, the more we understand love. That Love which was not found on Earth among humans, Humanity did not try to find without. They turned instead to their immediate surroundings. After millions of years we started to love the perishable. Some day we will learn to love the Imperishable. Jesus asked, “How can you learn to love the Eternal when you do not love humans?”

When disharmonious in every way, we like children, hit ourselves with our fists. Disharmony is like a fog or mist. Harmony is solid. When we try to hit disharmony, we hit Harmony. When disharmonious, we fight Harmony and Harmony hits back. We hurt ourselves against the Wall of Harmony. The Good hurts us back to teach us a lesson, otherwise we would always be disharmonious. In fighting the wall, the wall hurts us. We should try not to be disharmonious, because Harmony will hurt us.

God is Love, like a loving mother. How can a loving mother hurt her child? If a vicious child begins to hit its mother, the child's little hands are hurt. The child punishes itself by fighting that which it should love. If we strike against Love, we injure ourselves.

Really wrong should never be able to hurt us, but it does hurt us in those parts which are vulnerable. We should develop ourselves higher and make ourselves immune. When we are completely immune, Harmony cannot hurt us. If we are hurt, then there must be an element in us which can be hurt.

No one is truthful. We can try to be truthful. We often tell lies without knowing it. We could never tell a lie if there was not first the lie element in us. We should understand this and try to work out of it. We have the power to overcome. The good qualities are the dominating ones. If we are honest, we are with the Great Law. If we are in tune with the Great Law working through our Higher Self, we will win. However, we usually let mind interfere.

We have within us that something which continually wants to go higher. It is not our mind. Mind is earth-bound. If this were not so, we could be of no practical value on this earth plane. Today we are dominated by so-called practical things which are after all not very practical.

We cannot do something good to another and hurt ourselves by it. If it hurts us, there is something wrong with it.

In the Quest for the Unknown, we are seeking the Eternal. The Unknown is always facing us all of our life.

If a child would seek security, it would never leave the lap of its mother. The whole of Humanity wants security because we are still children. Security means limitations. We should not look for it as the main quest of life. Wild animals prefer the insecurity of open places, rather than the security of cages.

We have five gates or senses to penetrate into the Unknown. Inside of us, we correlate the outside impressions. Truth is both outside and inside. Truth demands completeness. Half a truth is a lie. Truth cannot be cut in half. It either is or is not. Security does not stimulate us, it ages us. The moment people feel secure, they begin to get old. Real peace of mind is an adventure. It gives us the opportunity to rise higher, and that is real peace of mind. The inner urge to go forward is adventure, the seeking of the Unknown. The more adventure, the more unfoldment there is. If a mistake is made, experience is gained. An inexperienced life is a worthless life. Experience is not in hiding in a safe place but in seeking adventure.

We are of an adventurous nature. The word adventure is often misunderstood. Most people imagine something reckless when they think of the word. People confuse words because they are not in tune with Harmony. Adventure means daring. We have the command, "Dare and Do." Daring is the mental attitude. Doing is the manifestation. This is usually physical because we are living on the physical plane at present. The whole of Creation is based on adventure. The Eternal is the Greatest adventurist because He dared to Create the Universe. He dared to Express.

The Cause is known by the Effect produced. If there were no Universe, there would be no Cause of the Universe, no such thing as the Eternal, the Absolute, the Great Father. The mere fact that the Eternal is called the Father of the Universe, shows that it means manifestation in the broadest sense of the word.

There is a distinct difference between daring and reckless. Reckless means unreasoned daring. Daring is reasoned. Life is the most daring thing in the Universe but it is not reckless. Mind dares to use its own prerogative to think. Truth dares to be Truthful. Love dares to be Harmonious.

Truth, since the beginning of the Great Adventure of Life, has been in the process of unfoldment. Now it is perhaps in its most conspicuous role of all time. Today the greatest test for all people is to be as truthful and accurate as possible. It is more difficult now than even a few years ago. Lies are becoming predominant in our lives.

Humans originally had to solve their problems as animals do, by instinct. Mental science claims it was subconsciousness. Such an estimate of our instinct is erroneous. Anything right cannot emanate from something that is fundamentally wrong. There is nothing harmonious in subconsciousness. Since millions of years we have tried to do right things, we have succeeded in weeding out of subconsciousness some disharmony. In the early days only our Higher Self could lead Humanity in the upward climb. At that time the Higher Self was like a Guardian Angel to Humanity in its infancy. Plants and animals try to do the best they can in their fight for life. This is not subconsciousness in them, it is instinct. The most powerful instinct of primitive humans was the desire to find the cause of existence. This seeking for the Unknown enabled them to stand the vicissitudes of that time. Primitive Humanity was very material. They had a great sense of material values. They valued the sun, the rain, even the storms because the storms cleared the atmosphere. All of this made them more and more unfold within themselves the Quest of the Unknown. When they saw the beautiful sun, they thought it was something supernatural. They began to adore it. There was probably born at that time, and for the first time, the human concept of God. No one knows what is meant by the word God. The Hindus call God "It" because no word can describe God. "It is still used to describe a person who has something undescribable. "It" is the Eternal. That which cannot be described. Humanity, though seeking for ages, still cannot understand "It." The Eternal defies any description.

When we cannot understand or reach something, it awakens adoration. It awakens Divinity. Familiarity breeds contempt. Familiarity is not loving someone or some thing. To be loving means to include a certain amount of respect. Familiarity stands for cheapness. Very familiar people are not truly loving. We should keep a distance between what we want to reach and ourselves. Life is nothing but a continual reaching and never reaching. If the horizons could be reached by walking, it would stop us like a wall. Therefore, it is very fortunate that the horizon always moves away from us.

Humans like monotony, yet monotony kills humans. Stagnation is death, the very reverse of movement, improvement, life. There is no "dare and do" in it. We are now passing through an unbelievable age of mediocrity. People instead of moving forward, are using their discoveries for comfort, which, in turn, is fostering laziness. Comfort is not Harmony. It can be harmonious, but it is only a part of Harmony, a momentary manifestation of Harmony. If comfort can be broken, it is not Harmony. We can violate Harmony but we can not break it. Violation is not breaking. A violated law is still there as a law. The only one who can break a law is the one who made it and then only by replacing it. The only one who can break Harmony is the Eternal. If He broke it, He would destroy Himself because the Eternal is Harmony.

Religions always put their Deities or statues in places where people could not reach them because of familiarization.

There is no reason for us to seek comfort. We should put Harmony of Life as a whole beyond temporary comfort. Comfort and Harmony, though going hand in hand, are just as different as is the human body and the Spirit dwelling in the body. We should always give predominance to Spirit. This is more practical now than ever before because we in our efforts to make life easy have confused the two. If Harmony and comfort were harmonious with each other, life would become very satisfactory. That is self evident. We should attend to the main things, and then add details if there is time.

We are living in a more comfortable world from a material viewpoint than perhaps ever before, but we are far from Harmony. There is nothing wrong with

Harmony. The wrong is on the side of comfort. Often we do not understand what real comfort is. We often achieve a temporary comfort which in the long run causes discomfort.

We should find out the price we pay for comfort. We pay for Harmony the price of Harmony, and for disharmony, the price of disharmony. Will the price be that of comfort or discomfort? The majority of people cannot see things right because they are in mental darkness. If we get real Harmony out of comfort, that is fine, but if we get disharmony, we should leave the comfort alone. The comfort of morphine is the price of agony. Everything disharmonious is the result of something which demands to be doped. We try in the present world situation to establish a bearable situation by using no end of dopes, such as trying to forget it or by using alcohol.

Life is not a picnic, but a continual struggle against disharmony. This has become so bad that people are beginning, on a large scale, to be afraid to hardly move in any direction. We should remember that obstacles are a great training and that they make us strong. There is a continual overcoming. We are never stationary. We cannot remain stationary. If we do not go forward, we are automatically pulled backward. Most people are non resistant because they are too lazy to be otherwise.

The United States in its desire for improvement has given the people physical comfort but the price we have paid is beyond the advantages.

The main object of life is to get happiness. We cannot buy happiness. We can buy comfort. Happiness requires a deep innate orientation. Harmony within is the key to all happiness. The Book of Life is written differently for each person. To copy life from someone else is as bad as to copy in school. In every direction we should exercise our sense of Harmony, but it should become automatic. We must continually train ourselves. The majority of people do not have a good time, because they are not trained in the sense of Harmony. We must fight outside disharmony and counteract it with inside Harmony. There is nothing more helpful than to be sincere. Insincerity is one of the greatest obstacles to coming in touch with Harmony. The world today has reached a most insincere condition. We can to some extent eliminate the troubles from ourselves. Try to synchronize as much a possible with the good things. There is nothing more important than enlightenment. More light means more joy and more Harmony, more life.

There have been several articles in newspapers recently about a Moral Rearmament program. The fundamental principle of this program is to live a life of sincerity, purity and love. Due to exceptionally clever advertising, this idea has spread all over the world. There are thousands of members with many centers throughout the world. Outstanding people have joined. This movement is a very good counter action to what is going on today. The Lightbearers, however, have something that is much better. Our FourSquare Principle says, "Be Strong, Be Wise, Be Honest, Be Loving." This is perfect. In the Moral Rearmament program there is something missing. Purity is very ambiguous as a word. It is little understood. Translated scientifically it means to be truthful, sincere, honest. People, usually physically inclined, consider it physically, as pure food, pure air, pure water, pure blood, etc. The highest purity is the Law. Law is ideal Purity. This is even noticeable on the physical plane. Pure metals, pure alloys, do no disintegrate. A pure body cannot be attacked by sickness. Our sicknesses are due to lack of 100% purity which permits the sickness to enter the loopholes. The saying, "The pure in heart shall see God" is true. Only when we are pure or sincere can we behold Harmony, and when we behold It, the Light of Harmony flows into us.

In Moral Rearmament, some details such as confessing sins publicly is wrong. Telling others of our mistakes will not help us. To tell to our self will help. Our mind should be the judge. Public confession produces a psychological effect on the individual and is somewhat helpful in that we throw off the burden and we feel relieved. This may work from a psychological point of view, but it is temporary and is not the way the Eternal works.

The only way to achieve results is to have courage. Stubbornness precludes courage. Courage is an intelligent attitude towards something wrong. Stubbornness is one of our greatest human handicaps. Fighting it is real courage, moral courage. Today everything sinks deeper and deeper into moral discouragement because of great stubbornness. To be unselfish is excellent, but it is only one of the ways that Love functions. Love is the total of all Harmony, and unselfishness is only a fraction.

Wisdom is the greatest gift for anyone, the Eternal included. Without wisdom there would be chaos. Without wisdom there is no possibility to reach Harmony. Wisdom is the raft on which we cross the River of Life from the shore of disharmony to the shore of Harmony. Wisdom is left out of Moral Rearmament because it, itself lacks wisdom.

In THE LIGHTBEARERS Teachings, we have how Nature works, and no one can exceed the Infinite, Eternal and all Powerful Nature. People attempt to tell the Great Law what should be done. It is sacrilege to tell God in prayers what God should do. The FourSquare Principle is not a prayer but a blueprint for life, which becomes valuable only when It is used. It contains nothing superfluous and there is nothing missing. Each part is in its own Corner, and each in logical order. It is the only way Humanity will reach perfection. It was Eternally that way and will be Eternally that way. The Lightbearers for the time being are a complete flop because we did not realize that on that platform alone we could have conquered the World. Universal Energy can be attacked as imagination, hypnotism, etc. There are no end of ways to doubt It because people have lost their childish simplicity when they trusted people.

Each one should be prepared to leave this World at a moment's notice. We should be as ready as is a bookkeeper to have his books examined. Each one is a bookkeeper of his own life. Wise people keep account of their book of life. The foolish imagine there will be no sudden call. The wise know that death can call suddenly. The fear which the majority has, is because unconsciously they feel the book of their life has not been kept properly. Each time we incarnate, we by and by learn to keep the book of life, sometimes through experience, sometimes through the help of others. This is a Law that is little understood by humans. Probably throughout the whole Universe there is bookkeeping. We cannot expect to reach perfection, but we should try to do our best. We should use courage and wisdom combined. We must be true and honest. Do not use alibis. I have tried to realize my mistakes and am ready to pass on.

Weakness is not permitted in Nature. Moral strength is more important than physical strength. We must be uncompromising for Truth. Whatever is wrong physically, mentally, emotionally within us, damages us. We should remember the FourSquare and try to express It as a pattern of life. No one in this World can criticize the FourSquare Principle. It will win and there is nothing mysterious about It. No religion can deny It, although they are denying Universal Energy. The FourSquare is beyond every criticism. Moral Rearmament is a feeble attempt to approach the FourSquare. The FourSquare is the Cornerstone which was placed in our hands and which we did not use in the building which we attempted to build. It is the greatest physical, mental and emotional rearmament which can come to this Earth.

The Flame which we use in our Assemblies is a small light burning in the oil of goodwill. If we observe it, it is a bright light. It is not how much we do but how well we do it. It is better to have a small bright light than a big dull light.

We should exercise our power of discrimination in every direction, and should learn to call things by their true name. Humanity has called things by their wrong names. That is why there is so much confusion. Clarity and strength should go hand in hand.

The minds of some people are sparkling or bright. Sparks sometimes turn into big flames. In prehistoric times when people first saw fire, probably from lightning, they tried to reproduce it by use of flints. The fire changed the whole history of mankind. It did it gradually through millions of years, but it succeeded, it conquered. THE LIGHTBEARER'S LIGHT is more than a spark. It is a steady light, one that endures, one from which we can get no end of sparks for more light. The passing of the light is not a mere ritual. It is a living reminder that we, in our minds, are desirous to pass the Light not only to our immediate friends, but to the whole of mankind. It is a reminder of Light shining in the darkness of human life.

Life as a whole on the Planet is far from being bright. We are continually reminded of it. We cannot close our eyes to that darkness. We should try to change it, not accept it. We should fight against it, if not by acts at least by thoughts. The more we realize it, the more we are conquering, not only for ourselves, but for our fellow prisoners on this Earth. We do not pass The Light for ourselves. It is for all Humanity, through a chain reaction.

There is nothing that is big to start with. Even planets and worlds started from a little nucleus. But the nucleus was strong. It was intensely active. It grew around itself, and out of it a world grew. It expressed life in an every-increasing way. It also expressed intelligence. Planets, animals and humans came into existence.

The FourSquare Principle is Eternal. There was never a time when it started. There will never be a time when it will cease. In The Flame is the creation of everything, the mission of everything, the purpose of life. In it there is Life, Wisdom, Law, Love. Everything tries to do things for itself. That is an inborn quality, not acquired, a foundation of everything, even of a little flame.

Curiosity is implanted in everything because we must notice everything. We must develop the power of observation. All life should learn in an ever increasing degree to stand on its own feet. A helping hand at the right time may help to solve the problem. There will never be a person born to this planet who will never tell a lie. Truth manifest is the Law. There has never been a time when people wanted more to be true than now, but never a time when there is less truth than now. We forget the highly Spiritual things of life due to the present ways of the World. The relations between people should be very simple, as they are in the rest of Creation. As time goes on in spite of the optimistic expectations of many people, things are by no means improving. THE LIGHTBEARERS have been prepared through warnings for many years. Unfortunately when one is prepared, when these things come, we cannot help being affected. It would not be human to be otherwise. We cannot be disconnected from Humanity. Each one of us gets the good and the bad impressions of Humanity through chain reactions.

Our individual mind is a part of Universal Mind, the same as our Higher Self is a part of Universal Intelligence. The sooner we realize these things, the better off we will be personally and collectively. Our human body shows that in spite of the diversity of different organs, they are still a part of one body, and must function harmoniously.

Everything in the World is doomed to failure unless the heart as well as the mind is in it. The two Eternal Twins work together for our good and for that of all Humanity. It is difficult, even for enlightened people, to understand how right feelings, thoughts and actions can affect the whole of Humanity. Unfortunately scientists disregard chain reactions in connection with atomic explosions. We should not forget that chain reaction is one of the Laws of Nature working on the physical, mental and spiritual planes. All Beings, even electrons receive Spiritual chain reactions, but most do not register them. Because of chain reactions, no thought, word, feeling or action is lost. The positive within us cannot be damaged because it has its origin in the Eternal. If it could be damaged, the Eternal would be damaged through chain reaction.

We have not exploited even one-millionth part of all the powers we have within us. We do not have enough confidence in ourselves. We do not need to have absolute confidence, but we should realize that there is an Eternal Center in the Universe and that we are in touch with It through chain reaction. We too much rely on what people say. Scientific minds are thinking today, but very little and very disharmoniously. When we base our thinking on limitations, this limits our ability to contact our Higher Self. The more we know of life the more we can handle life intelligently.

Nature is the uncompromising Truth. It does not lie. The reason we know so little of Nature is because Nature has so many facets. What we see now of Planets was once the truth. It is now only a record of what they were when the light left them millions of light years ago. The impressions, therefore, are not correct today. It is the same with our individual lives. Yesterday is past. We live only today. Past experience can be of great value to us, but we must live today.

The Universe is not expanding. Only our knowledge of it is expanding. In our quest for happiness, we are following the Laws of Nature. We have an increasing expansion of our realization of our own little universe within us. Harmony and happiness is within us. This has been taught by all great teachers. We do not need to look farther than to our own Higher Self. If we could realize this, not just talk about it, life would change for us, would be an ever-increasing happiness. The regeneration of wayward boys begins within. Their lower self is weakened by their Higher Self. The Higher Self wants to express Itself.

Science of Being is the foundation on which we can build a Tower reaching to Heaven. Its present conception by us and its manifestation is only the bottom layer.

When things go well it is easy to be optimistic and enjoy life. When things go wrong then things start to happen between those on the positive side and those on the negative side. All are on the positive side when things go well. No matter how many mistakes people make, their positive vibrations shield them for the time being from the negative causes they started. When things go wrong for a Nation or a group of individuals, there is an accumulation of negative vibrations. In every human field of life the attacker has the best chance to win. This is well recognized by military men.

There is nothing more perfect than Nature in Its Spiritual aspects. In It there is Eternal cooperation. Everything is synchronized.

The FourSquare Principle is the foundation of Creation, of our human life, of our Eternal life, of a plant, an animal, a stone, a planet. It was brought to Earth by THE LIGHTBEARERS in the best way that it has ever been brought. It is the most important part of our Teachings. Healing is only a side issue of the FourSquare. I realize it was a mistake in placing too much emphasis on healing and too little emphasis on the FourSquare. There never has been and never will be a mortal on this Earth who never makes a mistake. That which is built on a mistake, the sooner it collapses the better. True health can only endure on a strong foundation, and there is no stronger foundation than the FourSquare. The FourSquare which goes upward to the point of the pyramid, or Spirit, may also be projected downward to another point which is Matter. Spirit and Matter are opposite Poles of the same thing. Those who deny Matter see only half. They in reality also deny Spirit.

We should give with joy and receive with gratitude. This is a balanced condition. Most of us do not realize how far reaching the Law of Cause and Effect may be. When we do something, we should stop and ponder not only the immediate effect, but how far reaching the action may be. Most of life is lived in immediate effects. We can do things in a small way and in a big way. The big way is the FourSquare way. The small way is when we use only a part of the FourSquare. It seems almost unbelievable that we should use the big way in small things. People do not understand what is small and what is big. We judge things by volume and not by quality. Quantity is that which appears to us first, and quality next. Quantity is two-dimensional at best. Quality is three dimensional. Quantity is the small way. Quality is the big way.

We should see the True way, the deep way and not the superficial way. People who could have done better only looked at the superficial way. Because of this, the whole of history has been influenced. People who have been big on any plane have always sought quality first, and let quantity come later. The bean seed is a rather large seed. It produces a small plant and this lives only for one year. The plant produces only relatively few seeds and these are not of great value. In contrast, the Sequoia seed is a very small seed. It produces the largest known plant, and the plant lives for about 4000 years. It produces seeds, not only for one year, but for many, many years. It multiplies itself thousands and thousands and thousands of times, perhaps millions of times. We humans should always try to do things in a big way. The small way never produces good results. When the FourSquare Principle is not applied in life, it has unbelievably disastrous results.

In the true brotherhood of man, each one of us affects the whole of Humanity. There is no human being who can say he is standing alone. People are born equal before the Law. The Law affects each one in the FourSquare way. In this World there is a similarity of humans, but no equality. Equality does away with individuality. Equality when applied in a practical way completely destroys individuality. It is always the wisest policy to remain on good terms with one's immediate neighbors. If they do something wrong, just ignore them. A fight on the moral plane is justified but not on the physical plane. If we have the right to win, we invariably do so. When we do something the FourSquare way, we use wisdom, and wisdom in even a small way is best.

Humans today have paid so little attention to the Law of Vibration. There are three planes of Vibrations. The Spiritual is the highest, purest, most harmonious. In these we have our Being. We are born in them. They are our Higher Self, Ageless, Eternally, Harmonious. Mental Vibrations play a tremendous role in our Earthly existence. They are by far different from Spiritual Vibrations. They are a peculiar combination of right and wrong notes. Human life is fashioned on the mental plane, and then put into the physical plane. Mental vibrations are the seed, and physical vibrations are the soil. Physical vibrations are mostly harmonious in nature. When humans had no industrialization of the greatest part of mortal life, there were almost no disharmonious physical vibrations. Since industrialization has started, we have gone backwards. We have more disharmonious physical vibrations than ever before, such as noise, bad odors, etc.

They are so bad that people want to leave the cities and go to the county. The country is now being spoiled. Some vibrations have reached the point where they are almost unbearable. When this happens, humans usually coordinate their efforts to improve them. This is a very poor way and is due to their ignorance and laziness. These vibrations could have been stopped at the beginning with but little effort. We cannot do our best because we are so handicapped by outside vibrations. They cause us to feel downhearted, depressed and repressed. Love is the greatest redeeming power we have. Diseases and troubles of the physical body are manifestations of disharmonious vibrations within us. We should try to feel as harmonious as possible.

Universal Energy is the material symbol of the FourSquare.

Mental vibrations are stronger than physical vibrations, but are much more subtle. Those of us who know much more than the average person, know so little along this line that we are like children in the kindergarten. Humanity has by no means reached maturity. When they will reach it no one knows. There has been a greater growth along physical lines, and this has tended to make life pleasant. Mentally we have hardly scratched the surface as yet. Suppose that average people who have good intelligence have a knowledge of 1. The knowledge of very developed mental people with a knowledge of mental powers, manifestation, etc., would be 10. Those few who become mental guides of people, whose number is so limited that since historical times there have been only a few, their knowledge is 1,000, probably considerably more. Those who have advanced along mental lines very considerably have used only a fraction of what they have in them. They have developed only one millionth part of their mental powers. If they could use their powers, they could achieve physical manifestations.

The present condition of the World is probably the worst Humanity has ever seen on this Earth. Mentally, Humanity is in such a distressed condition that they do not know which way to turn. They are looking for leaders. People must rely on those who are guides in life. Such guides, if well informed, can do a great deal of good. If misinformed, they can do no end of evil. The trouble in the World is the responsibility of the leaders. A distorted or demoniac mental condition of the leaders produces all the trouble on this Planet.

The more mentally developed we are, the stronger are our thoughts. To develop our thoughts, we should be one-pointed. We should get to the point of what we want to say. One of our greatest shortcomings is that we waste no end of time in idle, unprofitable and destructive thinking. To be one-pointed means to concentrate our minds completely on the thing we do.

Life is more than just labor. It is also recreation, enjoyment, rest. Life should be balanced, so that we should not work in hours more than we devote to recreation. An individual with 5 or 6 hours of good mental work could achieve more than people formerly did in 11 or 12 hours. Life must be enjoyed. It can only be enjoyed if the enjoyment is wholehearted and to the point. Recreation should be one-pointed. If done in this way, we are never bored, never sophisticated. Sophisticated people in ancient Greece meant that Greece was beginning to decline. Sophistication shows that mentally we are on the decline. Sophisticated people cannot and do not enjoy anything. A more developed mental person will always understand those who are less developed. Concentration or one-pointedness, is a most practical thing. Life demands that we be one-pointed. If so, we will thoroughly enjoy life. Most people do not enjoy life. Each one is meant to live life the best we know how. We should not leave a stone unturned to achieve that. If we have trouble, we have not handled life properly. How can we learn to be one-pointed? We must have our mind concentrated on it. Then we grow mentally and harmoniously. We are now living in an age of mediocrity, a colorless, pale condition of life. This is due to a lack of concentration, or one-pointedness. If we are one-pointed, there must be and will be color. Very unpleasant things can still be colorful. We have mediocrity because everyone takes it easy in thinking and in feeling. We should combine intelligence with joy. We can do this if we are one-pointed.

Conditions as we see them today have always occurred when peoples and Nations became too self centered and took thing too easy. We should take life in an energetic way, but not in a hard way. Because we have let ourselves become decentered, we are mollycoddlers. A jellyfish is symbolic of something not interesting. They have no backbone. When thrown on the land, the sun rays act as a disintegrating power on their bodies. This explains graphically what human jellyfishes achieve in the wrong direction in life. They are often a jack of all trades. If tossed by the waves of life on the shore of the Beyond, they entirely disappear, so to speak. They leave no trace, no memory of themselves, no impressions. Not only their bodies dissolve, but their characters dissolve. They will be a jellyfish in the Beyond. They will have to pass through no end of incarnations as human jellyfish until they learn to have a backbone, a constructive activity of some kind.

We are born to have a constructive activity. The general attitude now is to take life easy, to not assume responsibility. Each person should capitalize on the qualities he is born with. What should we do? We cannot change the World. We should start form the inside. On the mental plane we have all the possibilities to improve ourselves. Thoughts have tremendous powers and should be developed.

Concentration is one of the powerful methods to improve thoughts. In the FourSquare we have simplified this procedure. We simply remember to live the FourSquare. This requires us to concentrate on the FourSquare. We have assurance that we will never be required to do more than we can actually do. This does not apply to things a person tries of his own direction to do. Becoming exhausted means that we do not have the ability to do the work, or we do not like it and fight inwardly against it.

The unfortunate part of life today is that thought vibrations of the whole of Humanity are disheartening and destructive. We have within us the power to resist any unpleasant vibration. It is a fine thing to be tolerant, but we must know how to be tolerant. We should be tolerant of individuals or groups who have a desire to improve and who have the possibility to improve.

The FourSquare Principle is a mirror in which we must see ourselves. The only stability we have is within ourselves, and in the FourSquare Principle. There is only one Truth. It is so big we will never know the whole of It. When we see, even through a little hole, the sun, we see the whole of the sun. This is symbolic of the FourSquare. We should lean on It, feel that in It we have a protection. It will never fail us. It is terrible to live in a world where we cannot trust or rely on anything. The Eternal will never fail us if we have faith in the Eternal. The FourSquare is not built by human hands but by the Divine Architect. It has Guided us since time immemorial. We should never surrender ourselves to any human or anything made by man. The only surrender should be to our Eternal Father. The Divine Model is within us, but is obscured by mind. The whole life of Humanity is based on Truth. People should realize their shortcomings and try to rise above them.

Wisdom belongs to the Higher Self, the Spiritual Ego. Wisdom tells us to remember the mistakes we made and try not to repeat them. At the bottom of despair there is a ray of hope. Only our human self can despair. The Higher Self knows no despair. The God of Love can never punish anyone. We punish ourselves. We have a chance to change ourselves. We should think of what our Higher Self tries to tell us, and not what our lower ego tries to shout at us. We are flowers in the Garden of the Eternal, each of us lovely and beautiful. Each one of us on this Earth plane is concerned with his own good, and then from his superabundance he should share with others. We see this even among wild animals.

A high rate of vibration is when we feel the very finest within us. Each is different. Each one of us has something very fine within us which is not asserting itself. People today have become so materially spellbound that they have forgotten idealism. The ideal to help physically, mentally and emotionally is on the right track, but unfortunately we turned away from our ideals. In real advancement, the ideals do not crumble, but become more and more lofty, broader and finer.

The human mind is set on destruction, but it must learn not to be that way. Humanity is rushing toward its own destruction because it has lost its ideal, its high rate of vibration. Truth is an ideal. Life can be so beautiful if we raise our rate of vibration. True love is the only unselfish power. Love is not affected by anything. It affects everything. There is within us an Eternal Phoenix which cannot be destroyed. We should say, "I trust my Father, I trust His Great Law." When we completely surrender to Him, we have no idea how it helps.

In our heart is the sacred chamber into which we can go and find there the strength and peace we need. The heart is the Altar of the Most High. Only the Eternal can enter. We are the priests who officiate at the Altar of the Eternal, in the Altar of our own heart.

We should try to conquer the downward movement of Polarity and Rhythm, and the disintegrating influences of our subconsciousness. Subconsciousness is now in a condition where it has to use all its powers in order to survive. Everyone fights to survive. No one is really ready ever to give up. Life is eternal motion. It is not supposed to stand still. We are supposed to be in perpetual motion.

Humans cannot build a machine which will never wear out. Why see the solution of the problem with our minds when it has been solved by the perpetual motion of the Universe? It has been solved not by the mind of man, but by the Intelligence of Nature. Science of Being solved the problem, because we know of the Energy of the Universe which never wears out. If Universal Energy were used to run a machine, the machine would not wear out, because Universal Energy preserves everything. A house in which harmonious people live lasts longer than an abandoned house.

If there had been no atomic bomb, there would have been a stop in the destruction by the human mind. The atomic bomb is now precipitating the whole

Humanity into a mental destruction. The bomb has already undermined the life of people. They live more and more in fear of what is going to happen. There has never been a greater world fear than now. We are now on the way to destruction. The greatest crime has been committed against Nature in Its Aspect of all Four Corners of the FourSquare.

Loyalty is the honor system put into practical operation. On the honor system everything on this Earth is based. No one can be patriotic to his Country on the basis of destruction. The correctness of the FourSquare cannot be denied. It is the only True Foundation.

We think we know so much that we do not need to remind ourselves of how little we know. The conscious self if properly guided, can conquer subconsciousness.

Innocence is very stupid. There is a difference between purity and innocence. We were born a combination of good and evil. Through successive incarnations we have had to learn what good is. Love is one of the surest and easiest ways to work it out. We have harmony with us. We are not products of disharmony. The present condition of the World, finds an echo within us. We should be as idealistic as possible. If we are not the Law of Evolution will do it for us. We should work with It rather than be forced to adjust ourselves to It. If we fall, we must get up and fight. We should say that no matter what is going to happen to us, we will try to make improvements. Do not let World fear affect us.

We pay now because we have mentally sown destruction. Let us each night before going to sleep, try to raise our minds to a higher rate of vibration. Think constructively for a few minutes before going to sleep. It will increase our optimistic attitude to the pessimistic world. If we try we may be spared in the great cataclysm.

In everything constructive, in our present conception of Time and Space, nothing in the material world is created at once. That which we call negative disregards time, but the positive does not. We say, give us time to grow big. If we feel we have time, we can do things well. Even the Great Law works slowly. The Mills of the great Law Grind Slowly, but Finely. Our human concept of time is given us as a protection. Quick people and slow people have different concepts of time. Each person has his own individual concept.

We are trespassers of the Law and are arraigned before the court of justice of our own self. A defense lawyer tries to gain time. At the end of our Earthly pilgrimage, most people ask for more time before passing on. We, in our ordinary daily life, ask the same thing. It is now more acute than ever. Give us time to make things right. We should make Wills when we are well and have the time to make them. There are people who seemingly never have the time to do anything. Others always have the time. They are not afraid that they have no time. They have confidence that the time is allowed to them, if they do the right thing at the right time.

The Law gives us the time to live up to the Blueprint which we carried to this Earth. This Blueprint was made by us, previous to this incarnation. We must try our best to live up to it. In the Blueprint, we have made our own time even for passing on from this plane. On the Clock of life, we have established the alarm. When the alarm strikes, the hour which we set, that is the call. A short life is not terrible. Whenever anyone dies we should not say that, "I wish they could have lived longer." This is a very deep and important truth. Never say that a passing on is a loss to Humanity. When our time has come, accept it. This idea is not new at all, but was known to the Ancients.

Life, in spite of all investigations is such a mystery, and death, in a way, is a still greater mystery. We can hear the Voice of Life, but we never hear the Voice of Death. We should say we have the time allotted to life. That is all there is to it. Doctors think they prolong life, but they are not able to do so.

There is an intelligent fatalism and a non-intelligent fatalism. Non-intelligent fatalism is where we give up too soon. Since we do not know the result, we should fight until we have done all we can, then leave it to the Great Law. The Great Law never makes a fool of a person. It never gives something to him, to take away from him. Let us take all things that come to us now which we cannot change, in a clam, serene and dignified way. Say it had to come, that the hour has struck. There is a certain purpose, a certain reason, a certain Law working back of everything that comes to us.

Life is not easy but we can make it acceptable. We can retire to our inner self and warm ourselves at the fire of our own heart. We can find there matches and a candle. What is a match? It is a human effort to make out of darkness, a light. When we strike a match and carry it where it belongs, we bear the light. The whole of life is made up of these little things, which if we do in the right way, will be credited to us. We can enter our inner self and close the door to outer life. All is within our inner self. It is our own home, our refuge. It has three sections, the basement floor on which we live, and the attic which we visit only occasionally, but which gives us the best view. In the basement of life, it is necessary to be strong. On the first floor are our daily activities. In the attic of life, is our Higher Self, which is the nearest to Heaven.

The Red Dragon or communism is the greatest danger Humanity ever had on this planet. No human can ever estimate the danger on the three planes. I know this danger because I started fighting it at the beginning of the Russian Revolution and have continued the fight ever since. I pleaded with President Woodrow Wilson and with members of his Cabinet. I lectured on its danger for a whole year. There were no results. I came to the conclusion that the Red Dragon is what Humanity brought upon itself. It was created by all of Humanity's shortcomings. We can rest assured that if it were not for the permission of the Great Law that Communism should win temporarily, it never would have won. Through my voice, Truth was spoken, but Humanity did not learn. Since the days of Jesus their hearts have been closed to God. Millions of people have been going to churches, but it was nothing but lip service. This is not my opinion. Facts have proven that their hearts were closed to the Eternal.

In my New Year's Message of several years ago, I explained the three adjustments. I said that in these days through which we are now passing. Humanity will call out to the Eternal and there will be no answer, because their hearts are closed to God. There are many evangelists now, but one moment of awakening is not enough. There must be a continual awakening. THE LIGHTBEARERS know God in His most glorious aspect as the Life of the Universe, as the Guiding Intelligence of the Universe, as the Law of Nature which governs the whole of the Universe, which Laws people can violate but never can break as they will always backfire, and, by Love, the Infinite One, the Eternal One which is proven by the Law of Attraction which keeps in harmony all celestial bodies, because they all bow to the Law. If there is a violation by a celestial body, the effect is annihilation. The Existence of the Eternal is not a belief, but a fact. This fact has to be realized on Earth, not momentarily, but as an earnest, continuing effort to do the right thing. Only the Eternal is Good and Right. Jesus called the Eternal his Father. He said "Only God is Good." He did not say that he, Jesus, was good. Mortals cannot expect to be perfect or right. Our understanding of right is nebulous, foggy. People look at right from a local, very narrow, personal viewpoint.

An effort, even if beclouded, will by and by lead to the right goal because a right effort comes from our Higher Self. When in our own self we begin to see some light and then concentrate our effort to reach that Light, we should go on even if prevented at times by others. When the faults are corrected, we will reach the Gates of harmony which no one can enter except by the FourSquare. It will take hundreds of thousands of incarnations to learn the whole lesson. We have been on this Planet perhaps thousands of millions of years, hundreds of millions since becoming human beings. Where are we? Even among the most advanced, we are not what we call good. There are no saints on this Earth, and never will be. This is just a human idea to excuse people of their own shortcomings, just an alibi.

Only THE LIGHTBEARERS, and they not with Universal Energy, but with the FourSquare Principle, can really help Humanity to reach the goal. It is the ultimate, the stamp of the Eternal, presented in the most simple way. Because we represent the FourSquare Principle, the greatest Power on Earth, the Red Dragon sees in us its worst danger. That is why we have to fight the greatest fight any organization ever fought. I realize that since I was chosen by the Eternal to bring to Humanity the FourSquare Principle and Science of Being, it is up to me to bear the first attack in every direction, both the impersonal evil and the very personal evil of Joseph Stalin and all the Communists. Some day we will understand and realize that the fight on this Planet is mainly a fight between the Communists and THE LIGHTBEARERS. There must be a leader on each side...Stalin and I, Svetozar. Stalin means man of steel. Svetozar means the one who bears the Light. Because I realize the magnitude of the fight and my great responsibility. I feel like in days gone by when David fought Goliath. On the outcome of that fight depended the future of Israel. Now the future of the whole of Humanity is depending upon the fight. If in that terrific fight, I will humanly perish, it will not mean anything, because it is my Higher Consciousness, the Ray of the Eternal which says to fight on, even beyond the gates of the Beyond. I cannot let any one of you take the lead in placing communists in the Sphere, because I am the attacker. You do not know the severity of an attack by the negative when it concentrates on one person.

The many accidents now are due to the evil of the world attacks. The ones who are weakest, are the first victims. Evil can only attack and conquer those who are weak. We must be Spiritually strong and fearless. If we conquer our fear in little things, then we can conquer in bigger things. Before we became humans we were fearless. Later we became fearful because we felt we had lost the support of the Eternal. The support is still there, but we are closed to It. Praying does not open us to It. Real prayer as defined by Jesus, is to do the Will of the Father. We do not need religions or any material structures. Just enter into our own sacred chamber, close the door and Commune with our God. That sacred chamber is found in the Higher Self which has never fallen from Its High Estate.

All contacts with Universal Energy are just imperfect ways to contact the Eternal in Its lowest form of energy. In our Higher Self, we contact the Eternal in Its Higher Aspects. Even the worst begin on earth as an Eternal Soul which will some day awaken, and then he will be able to join the throng which will be waiting for him to enter the Gates of Heaven. They are at present just open channels for evil to fight us. We must let the Power of the Sphere do the work.

Remember we have given to us that wonderful thing called Hope. If it were not for Hope, Humanity could not have endured all its tribulations. Primitive man was untaught in every direction, but he had the same characters we have now. They were full of all the undesirable traits of subconsciousness. They had no love in their human makeup, yet each had a Higher Self hidden in that human turmoil. It succeeded in working through. Now we have the same shortcomings, but more subtle, more of an undercurrent. We must fight it as primitive man did. First, there was a feeling that there is an Eternal Power greater than human power. Some found it in the murmur of trees, some in water, or in fire, or the sun, moon, stars. There has always been the search for the Eternal, the mysterious Unknown, the Father. That search was a mighty powerful instrument for unfolding human character. That search still goes on.

There should always be an increasing hope in us. The present conditions of the World are so tragic because the majority of people have lost their hope. Hope is the last string on the lyre which Humanity plays. When this string breaks it is hopeless. On the whole, it will never break. It will sustain Humanity. One of the ways the negative attacks humans is to cause them to lose hope. Old people usually have more hope. Young people today feel hopeless and consequently helpless. Sometimes it seems that hope can be destroyed because fear is blocking the way. Fear is the worst enemy on the physical, mental and emotional planes. There is no Spiritual fear because fear does not exist for Spirit. A really fearless individual does not know what the feeling of fear is, but there are very few of such individual. When fear comes, Hope is clogged up. Hope can never be really hit. It is only our mental concept of what Hope is. Hope is allied to Faith. It is an introduction to Faith, which is greater.

The reason people are so faithless now is because their sense of Hope is beclouded. Their fear, the great shadow in our life, plays the great trick on us, closing us to Hope. Hope is the beginning of expansion. In Hope we open ourselves. In fear we close ourselves. No Higher Power can pass though the body if the body does not react to It. The majority of people express themselves only 5%. This is one of the reasons why they do not enjoy themselves. An Electron is a celestial body, the smallest we know of. Hope and Faith have no place in a mechanically conceived Universe, which is the concept of most people. Humanity is more and more realizing that it has lost Hope. It is always trying to burn the incense of Hope by many means. Heaven or Harmony is based on Faith, the Absolute Faith that things will work out.

The motto of Hell is, "No Hope." This is the concept given to Humanity by Scientists. They give us more ease from a material view point, but also a mechanical World with no Hope. Yet, we cannot destroy within our own selves, our inborn Hope and Faith. We can apply what we learn from Nature to conquer our shortcomings, but we cannot conquer Nature. Nature is strong enough to strike back if Nature does not like what people do. The mechanical World promised Heaven but actually gave us Hell. In such a World we feel hopeless. When we touch nothing but the mechanical side of life, we cannot touch Hope, Faith, Love.

What can we do to remedy such a situation? We cannot create Hope. A thing created by Nature cannot be duplicated. All we have to do is remove the obstacle of fear. We not only can do this, but we have to do it. Say, "I shall not fear because I refuse to admit fear." This is the only way. Do not say that fear does not exist, because it does exist. We always have all the needed power to do the right thing.

The best care for old age is to be active, and the best activity is mental. We are the fruit of life and should ripen on the Tree of Life. We should have the sweetest meat in us when we reach the very end of our life. Let faith in our unfoldment and growth, give us the joy of life.

Loyalty is the thing which should guide every human. It is based on Hope and Faith. People are so disloyal today because they are living in a mechanical age. Once we are living in the mechanical age we should take the attitude of master of it, and not of slave. We think we cannot get along without the conveniences of the mechanical age. This is not so at all. Do not limit ourselves to it. This is easier to think and say than to act. Life in its final manifestation on the physical plane is by action.

When we admire something, it is the first step in turning our Being toward that which we admire. The more we admire, the closer we come to it. We should not permit material inventions to awaken within us admiration, because we will become subordinated and enslaved by them. If we want a thing strongly enough we will forget everything else to get it. People steal because they want the money that they steal. This is a primitive characteristic. Primitive people do not know the meaning of stealing. Children are the same. All people who act this way are immature in their minds. People with mature minds have reason controlling their desires. The others are tools of their desires.

Humans desires emanate from subconsciousness, and subconsciousness has no understanding of right and wrong. A child is a double Being, both right and wrong. We should permit the right Being to express itself, and not permit the wrong Being to express. Wanting to take something not belonging to a person, also shows an inferiority complex. If one expects something extraordinary from someone, we set in operation a certain law, and the individual is compelled, without knowing it, to live up to the law.

Among humans, who are not created by the Eternal, but are the product of our own minds, we are not born equal. Some are born with many talents, and some with hardly any. If we would accept this fact, we would not be so jealous. We should know that all comes to us is the direct effect of our own doing. We should always have the trust that it could be worse. This is a great moral uplift to have this trust that it could be worse. It gives us more Hope to fight. We should neither underrate or overrate ourselves. To do so shows an unbalanced condition and immaturity of mind.

We like to masquerade because we are born with a mask. Impersonation is a process of stealing. Actors and actresses are nothing but thieves in disguise. Nature tells us to be ourselves. Be true to ourselves, and we will be true to others, especially to the Eternal. The large number of actors and actresses at present indicates the fall of civilization. This was true in Rome when it fell. Hollywood is a cancer on the minds of people of the United States. That which is a pretense should not be taken as an inspiration by anyone. In no other county do we have such an adoration of actors and actresses. This indicates we are a Nation of thieves. We have a childish mentality.

The greatest aim of most people is to have fun. This again shows a childish mentality. Newspaper editors are so childish that they make almost a glorification of it. Life is no fun. It is a mighty serious thing for all of us. All people get is a fun of a life. This is different from having fun in life. The rabble are just naughty children. What shall we do? We are beset by enemies on every side. In this mechanical age we are like children, because we like a mechanical toy. The mechanical age kills our sense of life. We identify ourselves with mechanism. In addition to the deadening effects of the mechanical age, we are still mentally children. If we want to survive physically as well as mentally, we should try to break away mentally from the world in which we live. We must break our enslavement to it. If we do not, we will go under in the coming great cataclysm. Do not sponsor what is not essential in life, but sponsor the higher qualities in ourselves. Our inner strength will enable us to meet the situation. To break away was clearly indicated by Jesus when He said that we are in this World, but not of this World.

Mechanical gadgets should be our slaves. There is no need to accumulate more of them than we can use. Our initiative, our imagination is definitely greater than the mechanical devices. As long as we give so much attention to the mechanical side of life, we neglect the Spiritual side. We, as a civilization, have not improved people from a Spiritual viewpoint, and not even from the physical side. Mind is greater than the mechanical devices. We should assert our superiority to them. When people are mentally independent they can achieve something for themselves and for others. Mother Nature is greater than the mechanical age. She can shake her Body, and there is an earthquake. She can breathe heavily, and there is a tornado. She can perspire and there is excessive rain. What we can do is nothing to what Nature can do. If we free ourselves from enslavement, we will become something like humans and not robots. We get farther away from Nature, which is the Eternal Spirit of Harmony, of Balance. If we understand this, we will find we are in another World. Those who will not break away, will be crushed when the day of reckoning comes.

We, THE LIGHTBEARERS, and all others who think as we do, are now at the turning point in the right direction. We are turning toward a better condition. I feel it. There is no mechanical device as sensitive as a human being. Plants and animals are also very sensitive. Normalcy is only to be measured by our relation to the Laws and Forces of Nature. Some day we will realize what a Power is given us as Lightbearers. We know so little about Spiritual Values. We do not realize how one individual can represent the Spiritual Values of all Humanity. In spite of all difficulties, that which is right will conquer.

Unless an understanding is backed with energy, all other corners of the FourSquare are doomed to be a failure. Religions try to do the best they can, even though they put a false God before Humanity. They have miserably failed. We Lightbearers are in the front rank to fight for right, in the turning point which is here now. Where there is life, there is hope, and where there is hope there must be life. People with really fine qualities have the least energy. Gangsters have the most energy. The real touchstone of the value of one's character is to win in the great test. None of us has probably had our test yet.

Constructive work is the most enjoyable labor on this Planet. We can only enjoy life properly when we feel we have done something worthwhile. Energy is Eternal youth.

There is a youth more perennial than human action, and that is the youth of our minds and thoughts. In this turning point of World conditions, let us feel young. The body may get old, but mind should not. Let us look to the future and say that Eternal Youth, of which we are a part, will conquer.

There is an interval of time between the start of an event and its fulfillment. This turning point which we are now seeing, is deep within. It will be some time yet before its fulfillment. The outside of today, is the expression of the inside of yesterday. We are probably, as individuals, not feeling better because we have to express now all our past shortcomings. The old outworn should not be considered lost, but as a fertilizer for the future. No matter how hard or painful the experience, it is very useful. It shows us we are growing. When we have feelings of discouragement, sluggishness, sickness, if we can penetrate deeply, we discover the promise of a better condition in thc future. The evil now in the World is the evil of the past, perhaps thousands of years. These evil vibrations have accumulated to the point where we must get rid of them.

We have to get to our worst in order to bring out the best in us. The more discouraged, depressed, sick we will be, don't be discouraged. We cannot close our eyes to things, but we should know that the wrong in us must be exposed where it will die. It has to die. We should replace our older ideas, feelings and acts with better ones. If we do, we will find the way to happiness. We are now in a straight jacket. We feel the inner pressure of our own self. We should recognize the evil, face it and say that it is not going to keep me in a straight jacket. Do not say that the Eternal will come and liberate us.

Lesson Twenty

THE EGO

An Ego is the Self. There are two Selves to be considered. The Higher Self is the ruler of the lower self. The Higher Self is a Ray of the Great Eternal Self. It is the Individualized Projection of the Eternal, of the Supreme Intelligence, of that which has no name. Krishna called it THAT; That which is above all the Gods; That which does not describe anything, but includes everything. From THAT, Eternal Rays are continually projected. These are the Higher Self. The Higher Self, or Higher Ego, is spelled with a capital letter. The lower ego, or human self, with a small letter. They are the same thing, yet they are not.

They have the same relation to each other as a material object to its shadow. The object is substantial. The shadow has no substance. The object does not change. The shadow changes all the time, according to the light and the relation of the object to the light. The shadow is very faint the farther removed it is from the object. When directly under the light, there is no shadow. This is a very clear description of the Higher Ego and the human lower ego.

The lower ego is the adoptive child of the Higher Ego. It is the shadow of the Real Ego which is eternally connected with the Great Eternal Ego. It is the perverted image of the Higher Ego. Our small ego must be conquered. It should be held in abeyance. Our Higher Self has but very little influence on us at the present time, because we close our eyes to It.

Redemption is the journey toward that position where we are in complete relation to the Eternal Ego. This takes millions and millions of reincarnations. Most of these are back of us, but there are still many more, especially for the mass Humanity. As we are advancing in the process of Evolution, we are more and more freeing ourselves from our lower ego. When something is nearing its end, and feels it, it tries to fight for all it is worth to maintain its existence. Our subconsciousness is working that way. The fight is increasing in violence. Everything on this Planet has ample life to be sure it will win the fight. One thing only is doomed to lose and this is subconsciousness. It is condemned to lose. This is very fortunate, otherwise it would try to dominate this Earthly existence and The Beyond.

Almost for Eternity. Since Subconsciousness once started, it must come to an end.

Subconsciousness is unbelievably powerful, yet is so much less powerful than each of us, because we have in ourselves a Divine Ray of Intelligence which will endure throughout Eternity. We will in time be free from that terrible monster, that Frankenstein. Subconsciousness is seemingly growing stronger, because it is growing more violent. It is throwing out its latent viciousness. It is recently trying to create a great confusion. The majority of humans are little aware of the great fight between consciousness and subconsciousness. To win the fight we should use some fundamental rules. Our greatest enemy is fear. Fear is the great enemy also of animals, plants and probably of minerals. If it were not for fear, the mineral kingdom would probably endure forever. The diamond is fearless. No element can destroy a diamond so perfect is it in substance and structure. Only fire can destroy it, because it was born out of fire. A diamond can be burned making a most beautiful emerald green flame, of such a beautiful shade of green that it can be found in the rainbow. That is how it dies. All others die because they lack something which is due to some fear of which they are not aware.

Humans ooze fear. In primitive times, they were afraid in the material or physical aspect, as of big animals, etc. Even today physical fear seems to lurk everywhere, as in microbes, etc. Scientists instead of helping people to lose fear, have achieved the opposite. People have become more and more fearful, instead of less so. The mental health of the United States is below that of any other country in the World. Five percent are mentally sick. Criminals are insane or unbalanced people. There is insanity in almost all of life because of the influence of subconsciousness. Subconsciousness is the symbol of insanity. Without it people would feel that they can conquer whatever comes to them on the physical or mental planes. Subconsciousness is still trying to pour out its worse poison and Humanity accepts it. What can we do to counteract the terrible influence of the lower ego?

A projection is very easy to understand from the material point of view. There is a source of all vibrations, as those of sound, smell, sight, etc. These are all projections. They prove the cause. Otherwise we would not know the cause. Without vibrations or projections, there is no cause. From a mental viewpoint, thoughts are projections of the activities of the mind. When the mind goes blank, the brain for some reason ceases to project itself. Unless there are activities of the mind, there are no projections.

On the mental plane, if we do not feel love, we cannot project love. We can imitate it. Such vibrations, if they could be registered, would show that they were manufactured by the mind. There is a great injury to an individual if he represses love.

Radiations of love are so harmonious that the mere presence of such an individual produces harmony. Jesus was able to quiet the waves of the sea because He vibrated love, or harmony, all the time. The moment He realized mentally harmony, He could quiet storms and winds. The slightest manifestation of love is perfect, if it is really a manifestation of love.

Even manifestations of love must be proven. When we really love, we need no assurance. It cannot be NOT sure of itself. There can be no doubt. Love is the Law and the fulfillment of the Law. Love is inherent in us, but it must be cultivated, like the watering of a seed.

A Projection is simply a manifestation. There is more power in the word projection than in the world manifestation. As projection, our Higher Ego is Eternally living in the Eternal. Yet, since the Eternal has to project Itself out of Itself, it projects us. It can only project Itself into its own Self, which is Matter. Out of Spirit, the Great Self projects itself into Its own Self or Matter. If there were no Matter, there could be no manifestation, because it is a screen on which vibrations of Spirit, are registered. This is so subtle, so high, so refined that we can sense it better than we can understand it. We will never be able to understand Spirit. We will understand Matter more and more, but never completely. What we perceive through Matter is a manifestation of Spirit which we could never perceive otherwise.

We are concentrations of all Power, but in points. We receive vibrations from planets and stars, but we are not dependent on them. On Earth, we are the proof of the combined vibrations and Qualities of the Infinite. We do not know what is on other planets and stars.

Science of Being is the projection of the Mind of the Eternal, interpreted as well as can be by the human mind. I was chosen by the Eternal to be the channel through which the projection was made. The whole history of Humanity is in the dedication of the Textbook.

Our Higher Ego is essential for the existence of the Eternal Ego. Therefore, the Eternal would never leave us unprotected. The mere realization of this would make a great change in us. We are Gods, the children of the Great God. Our Ego is, was and forever will be. It cannot be destroyed. Even religions admit that the Ego is indestructible, Eternal. The Eternally Harmonious One cannot use disharmony to achieve Harmony. Why is it that the Eternal Ego is always ready to help, and yet we are so little aware of It and get so little help from It. This is an important question. It is based on a law, the Law of "Ask and Ye Shall Receive. "This is the Law of Demand

and Supply, of Cause and Effect. Ask properly and it is bound to materialize into an effect. The Law says it must be so. All Great Teachers have taught this. Mental scientists are now teaching the same thing. They insist on asking. They even make it stronger by repeating their insistence, under form of affirmations. There is a peculiar contradiction between affirming and asking. When we affirm, we affirm something that, has not yet taken place. If it had taken place there would be no need to affirm it. It is the conscious self that affirms that it is well. The subconsciousness knows that it is sick, because it is the cause of the sickness. The Higher Self knows that It is well. The conscious self knows that it is not well.

The Higher Self never lacks anything, but It never imposes Itself on the conscious Self. The Subconsciousness always imposes on consciousness. It always wants us to do something. The Superconsciousness always wants to be asked. Tactless people always tell things without being asked. When we ask we should always believe that we will receive. The Eternal must answer when asked. Asking is a challenge to the Eternal. But the Eternal can only answer questions. We cannot ask the Eternal direct. We ask through our Higher Self, which is the connecting link. It is the telephone. The Eternal projects Itself into our Higher Self, and not into our body. Universal Energy flows from the Eternal into our Higher Self.

Affirmations sometimes cause an effect on subconsciousness. If subconsciousness is sufficiently annoyed, it will accept. Subconsciousness swallows a negative statement, because subconsciousness is fundamentally negative. It is the Mother Lie. Jesus called subconsciousness, the devil or Satan. Religions usually make a mistake between Satan and Lucifer. Lucifer is the Symbol of Wisdom, and also of Truth. If Lucifer were a liar from the beginning, there could be no possible redemption for him. Lucifer is the Higher Self. Satan is subconsciousness. The satanic nature in us is that which is wrong in us. It is subconsciousness. Jesus never condemned Lucifer or the Higher Self.

Electrons are the bottom of materiality. Below them is Universal Energy. It is the first known to us, the first understanding of the Great Eternal. Scientists have reached an understanding of the Electrons. They are now ready for Universal Energy.

When the Higher Self reaches the materiel world, It crystallizes into a Human being. Subconsciousness, in which we move, live and have our being, is the world created by us. Not one spot in it can be called harmonious. There are two Creations. In the first the Eternal is the Creator. In the second, we are the creator.

The Eternal says, "We", we say "I". No one can stand alone. It cannot be done. The Eternal needs the assistance of Its own Creation and therefore says "We". Kings of old said, "We". The Pope of Rome and the Patriarchs of the Greek Orthodox Church say "We". The Eternal is the Great We. We are the "I" within the "We". Everything that is successful must work through collectively, but there must be a proper head. When we contact Universal Energy, we are no longer alone.

Believe when you ask. Doubt is like a shutter, an iron curtain. Faith clears the way. It should be an intelligent faith. The Eternal never does any destructive thing. In the Eternal Scheme there is no retribution, only reward. Retribution came into existence when we started subconsciousness. Without it, we would indulge forever in disharmony. That would be a black spot on the white Garment of the Eternal.

Honor is the highest sense we have of life. Since time immemorial, honor has been considered the thing to look up to. When we have learned enough our lessons, they will no longer come to us. What we have learned will become a part of us. Where we make a mistake, there we are weakest in our makeup, and there we pay the price. To be modest is a fine quality.

In the Silence of the Vibrations, we have the Eternal speaking to us. All vibrations have intelligence. The little intelligences in the vibrations of material objects, in rain, wind, etc., will respond if we ask them to do so. Jesus asked the water to support Him so He could walk on it. I using the same Principle can stand motionless in water with only my head and chest above water, my feet NOT touching bottom. I can also will myself to move in any direction while so standing.

We must revert to a mental simplicity. Those men who are most learned have almost a child like simplicity and attitude towards life. This does not mean to act childish. We should be as clean as we can mentally; that is remove as much as we can our fear, doubt, lying, etc. The FourSquare Principle helps in this more than anything else.

When we understand a thing, we realize it. Real understanding means realization, and realization means understanding. One cannot understand a thing without realizing it. There is a difference between mental and physical realization. Realization to start with is mental, but it extends itself to the physical. Complete understanding is the complete realization. Try to understand a thing first, then try mentally to make the understanding real, then try to apply realization in daily life. We will never be able to understand and realize fully the Eternal. If we could, we would be the Eternal. We would go from the effect back to the cause. The effect would die because of no cause.

Know thyself and thou shalt know all, is not quite true. We will never know All. When unfoldment reaches the end, it is the end. In the Eternal Scheme of Creation there is no beginning and no end. Unfoldment is based on understanding and realization. No one knows how unfoldment takes place in the Spiritual Realm, but it is probably in greater perfection of qualities. Everything in Nature is endowed with the Power of Observation, even plants. Flowers turn to the sun because they have observed that the sun is harmonious. When we have a four-dimensional state of consciousness, there will be no sequence in the Corners of the FourSquare. Now they appear to us in the order of Life, Mind, Truth, Love. The FourSquare is a Shield not made by human hands, not a theory, but by God Himself. Universal Energy is the FourSquare made a living unit in our life. Since we are so closed, only a little twinkle of It can enter our body, but we are not limited in our understanding of the FourSquare. We do not need to have any faith in It. Any individual can understand the Four Corners of the FourSquare.

The FourSquare Principle is so great that there is nothing else as great. It is the first and the last. There is no way to freedom, strength and happiness in life except through the FourSquare. The FourSquare is greater than the Teachings of Jesus on Love, because it is more understandable. Love, the Fourth Corner of the FourSquare, includes the first Three Corners.

Lesson Twenty-One

THE SIX POINTED STAR

Emblem of THE LIGHTBEARERS (design ©)

The Star for the New Age (Sixth Cycle of Humanity)

As the Morning Star presages the Sun, so this Star announces the New Day now about to dawn. The Star is six-pointed, symbolic of the Sixth Cycle.

At first sight this six-pointed figure appears as if made up of 2 interlaced triangles (the ancient Jewish symbol). However, whereas the 2 triangles of the Jews continue separate and are actually distinct from each, and never merge into each other, this new star is made up of one continuous band which interlaces within forming the appearance not only of 2 main triangles, but of many other triangles, each complete in itself, the total number of triangles being 20, symbolic of the 20th century.

The continuous single band symbolizes unity, the many triangles symbolize diversity.

If one traces this continuous band from any point, it will be found that there are six lines extending the full diameter of the figure, alternating with six shorter lines each the side of the two main triangles. The six diametric lines symbolize the complete expression of life and experience in the Sixth Cycle: the six shorter lines symbolize the as yet incomplete expression of the Sixth Cycle, which will find its completeness in the final unfoldment of the New Age.

The center of the figure is the sun of the New Age, which is both within the symbol and outside it, the rays of the sun extending eternally outwards into Cosmic Space.

SIX POINTED STAR

The Star of Balance of Wisdom was in ancient times called the Star of Bethlehem which rose above the horizon when Jesus was born, and according to tradition, the Magi brought their gifts to the new born Teacher. That Star, in a way, is our Star. More than ever that Wisdom is needed on this Earth.

Charity must always be exercised first at home. We are supposed to take care of our characters and wash away as many impurities as we can. If a strong, virulent thing strikes us, it is like black ink thrown on a white handkerchief. If we use a more virulent evil against it, we sometimes seem to succeed, but it destroys the fabric of our character. The holes thus formed are worse than the original black spot. The characters of most people are full of holes.

Lesson Twenty-Two

POINT OF POWER

The association of people with others has a tremendous effect on them. Only when they are strong enough not to be influenced by their surroundings, are they ready to take the next step, which is to try to influence the surroundings. It is not the position in the eyes of the World that makes a person a master. Money, education, knowledge, family ties are useful, but not essential. Our aim is to develop strength. We should not belittle ourselves by claiming we cannot do a certain thing. We cannot dictate to the Great Law. We should put that conceit into our pocket. If the Great Law demands that we do a certain thing, we are capable of doing it. The reason most people think they cannot do a certain thing is because they are under the influence of their subconsciousness. The bigger we are, the easier we can surmount mole hills. We should not consider any obstacle unsurmountable.

We must progress in life. We are never in a stable condition. We either progress or retrogress. If we are in a more or less satisfactory condition of health and energy we have not the right to say that we cannot do a thing. We should forget our weakness and try to develop the strength within us. There is nothing more satisfactory than to be able to say to ourselves that we overcame. Those who overcome will inherit the Earth, will receive a seat on the right hand of the Throne of Power. We all can do it but we must be strong. We should fight until the end, which is only the beginning of something better. We are born fighters. Every cell in our body is determined to succeed.

Should we, in mind, be less than the body? No, mind should be over the body. We have now to face the most difficult things Humanity ever faced. Let us be strong. What seems to be impossible for us, is possible, because it is imperative for the Great Law of our Eternal Father to flow through us. We must continue to fight. There are around us no end of constructive powers. Let us try to review in mind the things around us. Remember the kindnesses of those no longer with us who played an important role with us, whom we loved and who loved us. We dwell so much on the negative. Life around us is very disharmonious, even if it does not always touch us.

We must not live just an imaginary life, a life of wishes. Only that comes into our life for which we actually worked. A wish is a relief, in a psychological way, from frustration. We should dare and do. Since our conscious self is now not working properly, we should let our Higher Self take us by the hand and guide us. Trust that it is guiding us. We all have perceived real beauty in some form, in music, art, Nature, etc. Think of something that has come into our life that is so fine that it is almost Divine. These are Realities not wishes. Their vibrations are still somewhere and can be perceived. They are in our own subconsciousness to be resurrected by us. In subconsciousness the beautiful things are buried so deep they touch Superconsciousness. Superconsciousness is the beginning and the end. Subconsciousness is only a layer between the beginning and the end.

The evil becomes dust. It settles on everything and tries to hide from us the beauty. Let us not permit the dust of things undesirable to hide from us the things perennial. Do not try to find an excuse for wrong. Every cloud does not have a silver lining. Nature is not influenced by humans. It is what It is. The Laws of Nature call things by their right names. The more we try to explain something wrong, the more there is to explain. We find ourselves in a labyrinth. Right does not need to be explained. Things not right last a long time, but by and by they crumble into dust. In thinking of things beautiful we can outbalance those which are not. If we excuse wrong outside of us, we, without thinking, try to excuse the wrong within us. Only when a thought becomes an act can it bury itself into our subconsciousness. The Law of Nature accepts an act, either good or evil. We should see that an act is of the right kind. Not even death should stop us from doing something right. Life not lived properly is the greatest burden anyone can carry. There should be a desire to improve, then our Higher Self not only works with us, but reminds us to do a certain thing. If we are not reminded to do a certain thing, after reading or hearing about it, it is an indication that we have no desire to improve. Our Higher Self never refuses to help us if we honestly want to be helped. Progress is so slow because there is no sincere desire in us to really improve.

The more we advance into the darkness of the Unknown, the more we need a flame to illumine our path. When we reach a higher state of consciousness, we will not need a light. It is deep night today. We must use as much as we can of Light. We can increase our light, make it grow very bright, with our energy. Those who have reached a higher development of mind have the great advantage of feeling the freedom of lifting their minds into higher regions. They still need energy. We must develop within us more energy. We must be as active as we can physically, mentally and emotionally in the right direction.

If people did not have the quality of not accepting defeat, Humanity would long ago have been wiped off the Earth. We must carry this fight into The Beyond. If we do, we will wake up in The Beyond freed of that limitation. We can learn lessons here easier than when we will be disconnected from our physical bodies.

Lesson Twenty-Three

POINT OF ABUNDANCE

Today the sense of security is jeopardized. In the early days we had an inborn sense of security. There is now a desire to do something right to restore that sense of security. We are today reaching a most acute condition. If it had come suddenly none of us could have endured it. Out of the Compassion of the Eternal, we are stepping into it gradually. If it were not for the compassion of the Eternal we would never be able to endure what we did to ourselves. The Eternal feels in Its Heart how small we are, and how much the trouble means to us, and gives us the help we need, provided we are sincere. Sincerity sometimes cannot express itself because we block the door. Whatever happens now affects our minds and bodies, but not our Souls. If we turn to our Higher Self, It helps us to see the Power inside, the Ray of the Eternal, indissolubly connected with the Eternal. Our minds must realize this and then the barrier will be removed. The more we open our hearts to Abundance, the more It comes to us. We will get all we are capable of handling. We will be more and more open, if we realize that our Higher Self will never desert us. There is never a lack of material in the Storehouse of the Eternal.

Lesson Twenty-Four

POINT OF WISDOM

The Principle of THE LIGHTBEARERS Teachings is not revengeful, but we must be militant physically and mentally against all that is wrong. We usually try to find some excuse for wrong. If we do a wrong thing unintentionally it shows there is some wrong element in our character. If we would realize this we would unfold quicker than we do. If people would stick to the Principle of what is Right is Right, and what is wrong is wrong, one-half of the trouble of the World would be eliminated.

The Christian religion has been trying for nearly twenty centuries to teach people to rely on Higher Powers to save their Souls. If we have an immortal, Eternal Soul, which is an emanation of the Great Spirit, it does not need to be saved. What needs to be saved from its own self, is the human mind. It must be regenerated by using its own power. We should realize that in every step we take through the activity of our mind, our Eternal Father is always on our side. We have the support of the Great Law, but It is not pushing us. We must take the steps ourselves.

People who do not have higher development, who rely only on what they see, are helped through common sense. We should use common sense. Lack of common sense endangers all of our activities, and our whole life. Our mind is influenced by our subconsciousness which has no common sense. We are supposed to do everything we do with the help of God, in order to make everything harmonious. Stand for right, fight for right and never compromise. We do not have to accept mentally anything that is wrong. If we do something wrong and realize it, we throw the weight of the positive on it. We have mental powers, but we use them very little.

An historical personality, recognized by both religions and historians, King Solomon, when given his choice, chose Wisdom above wealth or a long, happy life. By that very choice he showed he was wise. He intuitively felt that Wisdom was the greatest of the three. Those of us who hesitate in giving an answer, are not inspired by our Higher Selves or the Eternal. We hesitate when we are confused by our human mind. King Solomon's answer was according to the Law. He knew enough not to make the Eternal wait. Because of his wise answer he received both the other choices. King Solomon is credited with having control over both this World and the Astral World. Wisdom should not be limited only to the visible World. We live simultaneously on two planes. We must exercise a great control over the invisible World, because it is directing the visible World. In the mental realm, or invisible World, all of our battles are fought, and won or lost. Those who have crossed into

The Beyond use unconsciously the Fourth Dimension in Matter. A material wall is no obstacle to them. Jesus was able to demonstrate the Fourth Dimension aspect especially after the Resurrection. Even today, some individuals can appear at will any place regardless of material walls. This indicates that these people have to some extent, unfolded their wisdom in the invisible realm. Because of the unbalanced condition of mind, very few people pay sufficient attention to both the visible and invisible Worlds.

To take the first step in the direction of Wisdom, we should drop excuses and alibis. Human Beings because they are not mature mentally, have to be treated as children. In dealing with children, a promise of a reward works beautifully. Prizes are now offered adults in very many activities. Mature people do not need rewards. There is nothing wrong with rewards. A crop is the reward of planting seeds. THE LIGHTBEARERS Commandment offers a reward, one for the present and not for the distant future or in The Beyond. From the practical point of view, this Commandment is greater than any Commandment so far given. It is the only Way. If we would only understand the value of It, we would not worry. *(See The Commandment of THE LIGHTBEARERS to the World on page 145.)*

The final aim justifies all efforts to solve a problem. All troubles of the World are due to the fact that people did not solve their problems. To do the right thing always pays ultimately. We must embody on the physical plane that which our mind has achieved through its unfoldment. If it were not for the material life, we would not be existing. Everything on this Earth, is material. Therefore we must give to Matter its proper due. If it were not for mind, Matter would be a formless substance. People are so little balanced that they see only one aspect of Creation. The Spiritual Realm, or Kingdom of Heaven, is within us. This Harmony could not manifest Itself without a material body, as long as we are on this plane. In The Beyond, Beings cannot achieve the fullness of their unfoldment, because they need the material world to do so. If they could, there would be no need to be incarnated again. If we come to this Earth, it is because we badly need it. We could not come without the full approval of the Great Law. Therefore all humans on this physical plane are more complete and superior to the inhabitants of The Beyond. Those in the Beyond have more apparent advantages, but these in reality are probably not advantages.

The main aim of the Great Law is to make us again perfect. The perfection can only be attained by the hard experiences of the material plane. The FourSquare Principle expressed in THE LIGHTBEARERS Commandment is the best way to achieve. We do not need any inspiration from The Beyond. Just turn our mental gaze toward our own human self, which is indissolubly connected with the Eternal. When we contact our Higher Self, we get a proper solution, and are not influenced by outside entities.

Every child has to learn the lesson of working through Matter. He cannot just learn mentally, as from books. The Ever-Present NOW for us is in the material plane. Always bear in mind wisdom. We all, as humans, make mistakes. We desire to be an optimist, but we overlook the fact that to be a true optimist is to think of the ultimate aim; that no matter if we do not know what to do, we should never say that it is the end. Where there is life, there is Hope. Hope is the last string on the Lyre of Life. If it were not for Hope, humanity could not have endured all the millions of years of its existence on Earth. Hope is among animals, plants and minerals, but it is not understood by them. The Law of the survival of the fittest, has hope back of it. Those who have hope are the ones who will survive. Hope is not Eternal, but it will last as long as we are incarnated on the physical plane. After that, hope will change into understanding…that condition where we know and therefore, we can. No one is really hopeless. That does not exist. Those who think they are, are the ones who are not strong enough to resist the pressure, who say what's the use to fight. Animals fight to the bitter end. We are better than animals.

This World seems more and more to be losing Hope. Hope is not what we call a thing tied to one particular condition. It is fundamental and not unreasonable. We often wish for unreasonable things, and then Hope against Hope. A thing started wrong can never end right. We should abandon the hope of making such a thing right. We should make a new start based on Truth and Honesty. We cannot compromise with wrong. All the wealth in the world cannot give us peace and harmony. We must be more consistently practical than ever before, because with the World collapsing, such people are the only ones fitted to survive, and are the only ones who will survive. Probably only a few will survive. Every human has the necessary Divine Spark which can be blown into a flame to survive.

Lesson Twenty-Five

POINT OF LAW OR TRUTH

Liberty is Truth. There is nothing more inspiring than Truth. We cannot have freedom without Truth, and cannot have Truth without freedom. We should make others feel that they cannot get away with a lie. They have no right to impose a lie on others. We do not need to submit ourselves to wrong. We must have faith to be strong enough to stand for Right. We are to work, not only for our own improvement, but for the whole of Humanity. If we help one, it benefits all.

We have not only our own puny little power to help us, but also the Great Law. When we do all we can, we should then leave it to the Great Law. If we understood this and had faith, there is no obstacle we could not overcome. No matter how many times we fail, let us still stand for Truth. Truth is bound to win. Truth and the Great Law are identical. We must have faith that Universal Truth is all around us, and that nothing can withstand the Power of that Law. We cannot disconnect the Great Law… Truth and Faith. Humanity as a whole, is a faithless generation on the higher plane. However, even atheists cannot do away with faith. They have translated faith to the material side of life. Something wrong will engulf Humanity, because of the faith of so many in Communism which is based on wrong principles. There has never been such an appalling and quick success as Communism.

Loyalty is based on Faith. Truth is the skeleton of everything. The pivot of the whole of Science of Being, and of life itself is Truth. When Truth is turned in the wrong direction, we have a lie. We live in a world of misplaced faith.

The Higher Self is the Representative on Earth of the Great Law. There has never been before a greater search for Truth. The search within us for Truth is so powerful that if we make a mistake, we are instantaneously reminded by our Higher Self. The more we vibrate to Truth, the more the great Law will help us, because we are open to It. Not one of us realizes the unbelievable Power and Value of Truth. Law and Truth are identical. Truth or Law is the pivot or axis on which the whole of life is revolving. Without It, there would be no harmonious functioning of Life, nothing but chaos. Cosmos means harmony. Chaos means disharmony. They can never blend.

Since human life started on the Planet, perhaps before, there has been a continual fight between law and lawlessness. The mind of primitive men was not clear enough to distinguish between good and evil. To learn this lesson, it has taken over a hundred million years. Throughout this time, Truth has been emerging out of the well. Truth is the pure water in abundance at the bottom of a deep well. The Laws of Nature are the functioning of Truth.

We have already entered the Cycle of Truth. Strangely, however, there has never before been such a wholesale expression of lies. We can judge that which is good, by that which is evil. It is easier to see that which is evil than that which is good. There has been more display of evil since the beginning of the First World War in 1914, than ever before. It is growing worse. The unbelievable distribution and manifestation of evil now shows that the Laws of Nature are claiming their own. The hidden evil is coming to the surface by the Power of Truth. Nothing is hopeless.

No mortal can create good. It is created by the Eternal. Mortals create evil. We have enough understanding of Truth now to fight successfully and win. The Divine Power was not given to us to voice a lie. We have a wonderful chance now to side with Truth. Truth is based on common sense. Do not be depressed by conditions today. It is a physical, mental and emotion housecleaning. When we think of results that will be obtained, it is a wonderful thing. We must today be more watchful because there are more pitfalls than ever before.

We are instinctively truthful beings. Instinct is a Spiritual quality. People think it is an animal quality, because they have observed it in animals. Animals do instinctively feel things. Dogs perceive a danger which they cannot see, hear or smell. Plants, minerals, gases also have instinct. Humans have lost this primary instinct. Something which Nature gave us cannot be actually lost, but anything physical, mental or Spiritual which is not used, shrivels. We have within us a sleeping giant, which is those mental and Spiritual organs which are dormant, due to dis-use. These can be awakened. This is done usually through Truth. Jesus said, "You shall know the Truth, and the Truth shall set you free." The more we awaken our sense of Truth, the more we become free. At present this is more difficult to do than ever before. We have a world which wants to imitate the world created by Nature. Our present world is a synthetic world. Nature created everything for the use of man. Science tries to do better than Nature, and in our own eyes, they succeed. What they produce does not last. Every imitation is a lie. Truth cannot be imitated. IT IS.

Today we are living in a world of greater pretense than ever before. Genuine things have roots in Nature which are Eternal, indestructible. The more we live in the scientific age, the more we are discovering, in a way, Truth, and also going farther away from Truth, in another respect. Science should discover the things which Nature has in store, which are almost limitless. The more we remain in tune with Nature, the more Nature reveals Itself.

Humans have created a mechanical age. This makes our mind mechanical also. This deadens our minds because mechanical things are not living things. The most wonderful device created is the human body. Scientists stand in admiration before their clumsy creations and do not give credit to Nature. They usually think they know everything. The great scientists do not take this view. They are very modest.

The problem now is to discriminate between what is right and what is wrong. We are not living now according to our instinct, which is our Higher Self. If humans would grant that all younger parts of creation have Souls, the whole concept of Creation would be changed. A piece of wood would become dust, if its Soul left. People today are very unaware of Truth. Each time we tell a lie, we disconnect ourselves from Truth. The size of a lie does not make any difference. We believe blindly what others say and do not use our discrimination. The more we discriminate, the more we are in tune with Truth. We have all the elements of Truth within us. Our lack of understanding this prevents us from siding with that which is inside. Truth always wins in the end. Why don't we start with that which wins in the end? If we would be more trusting of those fine qualities within us, we would be the winner. In spite of all our desires, we are not yet ready to do it. Truth never destroys anything. It is fundamentally a creative Power. Today is Truth. Tomorrow is a belief in something.

Lesson Twenty-Six

POINT OF HARMONY

This life is full of paradoxes, as for example that of strength and love. Where strength is shown, love is back of it. Love is supposed to be gentle. Humans unfortunately reason, but very little. If they would pay attention to the School of Life and of Nature, they would learn more than in schools and colleges.

The Soul of the worst criminal is beautiful. Our Higher Self is the Mind of Spirit, of a son of God, not a fallen Spirit, not an exile from Harmony, but one who is continually beholding the Eternal. What humans have of Love today, is so little, so unimportant, so misunderstood, so misinterpreted that it is not worth considering. Love holds the whole Universe together. Love never forgives. It always helps. It says, "Unless you do the right thing, you will have plenty of trouble." If we forgive the human mind, the worse it will grow. Love is unbelievable Strength.

In this particular age, we have to face one of the greatest explosions of hatred we have ever known. It is really hatred mixed with jealousy. People hate that of which they are jealous. The moment we are jealous, we start to hate. We cannot be jealous without hating, but we can hate without being jealous. We have some excuse if we hate something wrong, but really we should, instead impassionately fight it. Machines love but do not hate. If a machine is properly treated, it gives its best, that is, it loves. A machine is built on love, based on harmony. Humans usually permit the machine to deteriorate. We have permitted ourselves to become inferior to machines, when we should be superior to them.

We are born to this world to build our character. All else is secondary. Love is the greatest discipline. We cannot build it with hatred. Self discipline means to make ones self harmonious. When we learn self discipline, we learn Love. Law has the Key to Love. Love does not have the Key to Law. Real gentleness needs great self control. Unfortunately humans, when they speak of Love, think of something of an emotional nature.

This is the best they know. They judge Love from a material point of view. Emotions are material. They are something like explosions. They are safety valves, comparable to volcanoes. If it were not for volcanoes, and probably also earthquakes, there might have been an explosion of the Earth's crust causing perhaps the loss of all life on Earth. When too much pressure accumulates in the heart, which is the heart of our subconsciousness, the pressure manifests as human emotions. All human emotions are subject to decline but at times something higher springs up, coming from harmony within us, which is the Law of Eternal Harmony. Humanity without knowing it tries to find this Harmony within, that is, to satisfy their inner yearning, but they try through material channels. They usually never find it.

The most usual method is by mating, then through association with people, some by association with animals. Some try to find real Love in plants and in beautiful scenery, some in a hobby or by collecting things. All of these are only a feeble reflection of real Love. The Ancients discovered that the best way to satisfy their inner yearning, is to turn to the source of all Love, to the Eternal. It is a far away journey and but few are willing to take the journey. If humans would be honest and loyal to any of those minor avenues, they would find the goal. All roads lead to the Eternal, even if some are crooked roads.

The only happiness on Earth is when we behold the real good. Everything seeks happiness. Most people seek it in the green fields far away, often in those belonging to someone else. We can only find happiness within our own selves. There we will find the answer to our quest. No one can ever enter into the inner sanctuary of another, because there are invisible barriers. If one tries, something stops him. It always ends in frustration, and it will be so throughout Eternity. If one could, he would displace the rightful ruler and would remain there. It would be like a thief entering a house and proclaiming himself ruler of the house.

When we have found the Kingdom of Heaven, we will emerge out of darkness. Then we will begin to express True Love. All previous concepts of Love will seem childish. Humanity is seeking that inner balance, which they cannot find in the outside world because their mind is a dark lens, a cataract through which they do not see things as they are. We are further and further removing ourselves from our original condition as children of Nature. Humans want to improve on Nature, and in their own opinion, they think they succeed. We are all heirs to the Harmony of Nature.

People try to remove the mental cataract, but instead of a clearer vision, they get more confusion. We can only see properly if we have our natural mental vision. If we see life only through man-made help, we have only partial relief. The cataract is our own shortcomings, a stone wall between hour Higher Self and outer Harmony. What can dissolve the mental cataract? Harmony is able to remove the wall. The foundation of Harmony is Love. It is within us. It is more precious than anything else on Earth. It is not only Peace but also Power, the overwhelming Power which, when awakened, is always with us. If we develop our latent powers of Spirituality and

Harmony, it will save us. This can never be done by proxy. One has to work for it day by day. Humans do not want to make the effort. They want a shortcut.

Perfection must be solved. It is a "Pearl of Great Price." What is hopeless for us to achieve, the Eternal can do for us, but the preparation for it may take many years. Mental healing is usually not instantaneous because it is opposed by the subconsciousness. The easiest effort is the FourSquare effort. There is no easier or surer way. The FourSquare is the Symbol of Harmony and a means to Harmony. This is our problem, let us make it our foremost aim to bring from within us that Harmony. We must not be satisfied to keep on the same level. We must increase. The destructive side of life unfortunately is increasing by leaps and bounds, and we must increase the constructive side in the same way.

Love remembers. It never forgets. One cannot love without remembering. We cannot forget what someone does for us if we love that person. Temporary flare ups are not real Love. A devoted dog shows a spiritual Love to its master. This is much higher than the ordinary concept of love by humans. The ordinary concept of love is seated in the subconsciousness of individuals. Originally there was no love in subconsciousness. Later it penetrated down into subconsciousness. To fall in love means to go down in love. Those who really love do not fall in love, they love.

There is no love without respect. We cannot love someone whom we do not respect. Love is by nature a protection. Love coming from the Higher Source never makes a mistake, but when it comes from the cesspool of subconsciousness, it takes a different form. Because of the lack of True Love, we have fallen below animals. We will go on down to rock bottom, then people will call out for help. We only discover true friends in moments of adversity. There is now an unbelievable lack of True Love. We should try to bring out the best in us. We are born aristocrats, the children of the King of Kings. We are born Princes of the finest blue blood in the Universe. We now forget our true nature because we are so lowered by our everyday life, and our contacts with others. Peace will come when that which stirs up trouble within us, will subside. This misguidance is from subconsciousness. Let us try to find True Guidance from the Higher Self within us.

Lesson Twenty-Seven

POINT OF PROTECTION

Protection should be greater as evil increases. The Power within us, supported by the Power without, is greater than all the trouble. Scientists do not know what the Power protecting the Universe is, but THE LIGHTBEARERS know much more than other people. This Power is closer than the air we breathe. It makes a contact with the Power or Life within us. We cannot expect that suddenly we will become changed beings, but each day we change a little. We are now confronted by the greatest power of destructiveness, but that which is constructive must win. Only our lack of faith prevents us from this realization. That Power never fails. If for one instant the Laws of the Universe would fail, there would be chaos in the Cosmos.

Our conscious self is so affected by our subconsciousness that we have a continual battle. The World is now divided between the Higher Self and the lower self. We must choose on which side we will be. We are human and are all the time vacillating due to the Law of Rhythm. No one can ever avoid the Law of Retribution. THE LIGHTBEARERS have a great advantage in comparison to others. Help comes from unexpected sources, when we do the right thing, thus starting right causes. Ignorance is our worst enemy. The World is a mixture of good and evil. If we deny evil, we bring about a blindness to evil.

The most important knowledge we can acquire, is to learn how to take care of our business of living. We should learn not to destroy the natural balance within us. The Spiritual Concept includes the proper handling of Life.

The Commandment of THE LIGHTBEARERS to the World

"BE MAN"—Express in every act of yours All ENERGY, INTELLIGENCE, TRUTH and LOVE; thus Living only will you live; thus acting only can you build to Freedom, Strength and Happiness in Life. This is your Problem: be this your Foremost Aim.

The Commandment! Throughout the whole of Eternity there will never be another Commandment given. THE LIGHTBEARERS Commandment is that great. The FourSquare is the essence of Truth. It covers the ground so completely, that even God, if He wanted to improve It, He could not do so. It was the foundation of Creation, throughout all Eternity, and will continue so for Eternity. The Infinite Intelligence opened my eyes to see the logical sequence of the Four Corners. If we have faith that the FourSquare is all there is, we are open to the Eternal. If we have no faith, we close the door to the Eternal.

~Eugene Fersen ~

“Vibrations”

Father, I am Thy Individualized Projection into Thine Own Eternal substance proceeding from Thee, indissolubly connected with Thee, manifesting all Thy Qualities and Powers. I am indeed the image and likeness of Thee.

Mother, I am Thy Individualized Projection into Thine Own Eternal substance proceeding from Thee, indissolubly connected with Thee, manifesting all Thy Qualities and Powers. I am indeed the image and likeness of Thee.

Great Eternal One — The Creator of the Universe, I Thy created am Thy Individualized Projection into Thine Own Eternal substance proceeding from Thee, indissolubly connected with Thee, manifesting all They Qualities and Powers. I am indeed the image and likeness of Thee.

Divine Mother, Love Eternal, I, Thy child am Thy Individualized Projection into Thine Own Eternal substance proceeding from Thee, indissolubly connected with Thee, manifesting all Thy Qualities and Powers. I am indeed the image and likeness of Thee.

And, as such, I am open on the physical plane to the influx of Divine Wisdom, Infinite Abundance and Limitless Supply.

And, on the Mental Plane my mind is open to the influx of Divine Wisdom and Thy Limitless Supply.

And on the Plane of Spirit forever is my Soul open to the influx of Divine Wisdom, Eternal Love and Limitless Supply.

And this realization comes down to the Mental Plane.

And on down to my everyday life where I KNOW I am open to the influx of Divine Wisdom, Divine Abundance and Limitless Supply.

Father, Thou are inspiring me on the Spiritual Plane. Thou art guiding me on the Mental Plane. Thou art sustaining and protecting me on the Physical Plane.

Made in the USA
Charleston, SC
12 September 2011